Who Will Feed China?

OTHER NORTON/WORLDWATCH BOOKS

Lester R. Brown et al.

State of the World 1984
State of the World 1985
State of the World 1986
State of the World 1987
State of the World 1988
State of the World 1989
State of the World 1990
State of the World 1991
State of the World 1992
State of the World 1993
State of the World 1994
State of the World 1995
Vital Signs 1992
Vital Signs 1993
Vital Signs 1994
Vital Signs 1995

ENVIRONMENTAL ALERT SERIES

Lester R. Brown et al.
Saving the Planet

Alan Thein Durning
How Much Is Enough?

Sandra Postel
Last Oasis

Lester R. Brown
Hal Kane
Full House

Christopher Flavin
Nicholas Lenssen
Power Surge

WHO WILL FEED CHINA?

Wake-Up Call for a Small Planet

Lester R. Brown

The Worldwatch Environmental Alert Series
Linda Starke, Series Editor

W · W · NORTON & COMPANY
NEW YORK   LONDON

First Edition

The text of this book is composed in Plantin
with the display set in Zapf Book Medium.
Composition and Manufacturing by the Haddon Craftsmen, Inc.

ISBN 0-393-03897-1

ISBN 0-393-31409-X (pbk)

W.W. Norton & Company, Inc., 500 Fifth Avenue, New York, N.Y. 10110
W.W. Norton & Company Ltd., 10 Coptic Street, London WC1A 1PU

1 2 3 4 5 6 7 8 9 0

This book is printed on recycled paper

Contents

Acknowledgments

At the outset, I particularly want to note the contribution of my executive assistant, Reah Janise Kauffman, who helped produce one draft after another in rapid succession. Her enthusiasm for the project was a constant source of support.

Our series editor, Linda Starke, who also edits our two annual books—*State of the World* and *Vital Signs*—helped get rid of some of the repetition in early drafts, sharpened the thinking, and clarified the writing. She made a demanding schedule seem easy. At W.W. Norton, Iva Ashner and Andrew Marasia made such a schedule possible.

Elena Wilken helped with the research for the book. She was especially helpful in compiling and analyzing the data used in the analysis, only a small part of which actually appears in the book itself. Coming down the homestretch, Anjali Acharya helped with the final review and last minute fact-checking.

Several of my colleagues read the manuscript in various stages and made helpful comments. Among them are Derek Denniston, Sandra Postel, David Roodman, and Aaron Sachs. Each gave me suggestions that helped shape the final product.

Reviewers from outside the Institute included Dana Dalrymple, agriculture specialist with the Agency for International Development (AID). In reviewing this manuscript and providing detailed suggestions, Dana maintained a longstanding tradition of providing helpful commentary on my manuscripts that began nearly 30 years ago. Laurie Burnham, former editor with *Scientific American* and now editor of Shearwater books, a division of Island Press, made several useful suggestions in dealing with some of the broad questions raised by the analysis. Carl Haub of the Population Reference Bureau provided his usual useful review of population numbers. Paul Hawkins, business leader and environmentalist, interrupted a busy schedule to provide quick feedback.

In doing this analysis, I drew heavily on the world agriculture database from the U.S. Department of Agriculture (USDA). It is invaluable, as are the reports of the veteran team of China analysts at the department, including W. Hunter Colby, Frederick W. Crook, and Francis C. Tuan.

We are indebted to the Wallace Genetic Foundation, which provided financial support for *Full House: Reassessing the Earth's Population Carrying Capacity*, the book that spawned this project. The foundation's three trustees—Robert B. Wallace, Jean Wallace Douglas, and Henry B. Wallace—all have a longstanding interest in food and population. Others that support the Institute's work include the Nathan Cummings, Geraldine R. Dodge, Ford, W. Alton Jones, John D. and Catherine T. MacArthur, Andrew W. Mellon, Edward John Noble, Surdna, Turner, and Weeden foundations; the Pew Memorial Trust; and the Prickett, Rockefeller Brothers, and the United Nations Population funds.

A final debt of long standing that I would like to acknowledge is to the USDA management team that was so helpful to me early in my career. When I joined the Department's Foreign Agricultural Service on June 1, 1959, I had the great fortune to work in the Far East and South Asia branch of the Regional Analysis Division, which was headed by Quentin West. Quentin never gave me the feeling that the system was putting any limits on what I could do; the only limits were those imposed by my own intellect and energy level.

In the years since, most of them spent in management, I have tried to create a similar environment for my own colleagues. To the extent that I have been successful in creating an environment where young people could unfold at Worldwatch, it is due in no small part to the early influence of Quentin West.

Division director Wilhelm Anderson set the quality standard for our research, both initially in the Foreign Agricultural Service and later when the division was transferred to the Economic Research Service (ERS) at the beginning of the Kennedy administration. At the helm of ERS was Nathan Koffsky, also a talented administrator, whom I was privileged to work with as special assistant before moving to the Secretary's office. As a member of Secretary Orville Freeman's Staff Economist Group, I worked closely with John Schnittker, Under-Secretary, and developed an appreciation for his incisive thinking.

Helping to guide me in those early years was Joe Robertson, Assistant Secretary for Administration. It was largely because of him that I went from an entry level GS-7 position in 1959 to Administrator in 1966 of the International Agricultural Development Service, managing technical assistance programs in some 40 countries. Joe's goal always was to challenge employees to do more than they thought they were capable of. This, too, has been integrated into my own management philosophy.

And during these years, I got to know Secretary Orville Freeman, as both a professional colleague and a friend. This close friendship and working relationship has continued until today, with Freeman serving as chairman of the Worldwatch Institute Board of Directors throughout its entire 20 years.

The other person from whom I learned about leadership and commitment was the late James P. Grant. After working together to introduce the Green Revolution in Turkey, where Jim was the AID mission director, I accepted Jim's invitation to help him start the Overseas Development Council. From Jim, who later went on to provide UNICEF with some 16 years of spectacular leadership, I learned a great deal about how things worked in the nonprofit sector. It was the beginning of 26 rewarding years spent on Massachusetts Avenue's "think tank row."

Lester R. Brown

Worldwatch Institute
1776 Massachusetts Ave., N.W.
Washington, D.C. 20036

June 1995

Editor's Note

This is the sixth book in the Environmental Alert Series, and the first to deal with a single country. But as Lester Brown points out here, China is such a large nation—in terms of both population and economy—that its successes and failures affect us all. As he notes in Chapter 1, in an integrated world economy, rising food prices in China will translate into rising food prices everywhere. Land scarcity there will become everyone's land scarcity. And water scarcity in China will affect the entire world.

Who Will Feed China? does not aim to point a finger at just one country and say future world security rests on decisions taken there. Rather, the goal of looking at China's food prospects for the next several decades is to highlight the unsustainability of most national economic and population trends. And to provide a "wake-up call" for political leaders and the public alike so that we can address these issues before the worst-case scenario comes to pass.

One of the more interesting points Lester raises is the impact of the 1959–61 famine in China on the national psyche and the strategic thinking of political leaders there. The deaths of 30 million Chinese, and in such a short time, understandably affected public thinking about food security. But that memory—and the fear it gives rise to concerning a dependence on outsiders for food—should not be allowed to cloud leaders' perception of the reality facing China today. As the nation continues to industrialize, it will need to import grain to meet rising consumer expectations and the needs of growing numbers. By the same token, leaders outside China need to face the reality of how this could affect food prices in their countries.

Since 1991, the Environmental Alert Series has dealt with environmentally sustainable economic development, consumerism and the search for sufficiency in our life-styles, water scarcity and the need to invest in efficiency in that sector, pressure on food production systems worldwide, and the coming energy revolution. (See list of titles on page 2.) We see these short volumes on specific topics as complementing the Institute's two annual volumes, *State of the World* and *Vital Signs*, and hope that you find them useful in your own efforts to build a sustainable world.

Linda Starke, Series Editor

Foreword

My concern with China's long-term food prospect first arose in 1988 while I was reading the World Grain Database prepared by the U.S. Department of Agriculture. This remarkably useful resource contains the area, yield, and production of each grain in every country from 1950 forward. I may have been one of the few readers to notice that if countries become densely populated before they industrialize, they inevitably suffer a heavy loss of cropland.

If industrialization is rapid, the loss of cropland quickly overrides the rise in land productivity, leading to a decline in grain production. The same industrialization that shrinks the cropland area also raises income, and with it the consumption of livestock products and the demand for grain. Ironically, the faster industrialization proceeds, the more rapidly the gap widens between rising demand and falling production.

Before China, only three countries—Japan, South Korea, and Taiwan—were densely populated before they industrialized. Within 30 years or so, each had gone from being largely self-sufficient in grain to importing most of their supplies. In 1994, Japan imported 72 percent of the grain it consumed, for South Korea the figure was 66 percent, and for Taiwan, 76 percent. In none of these countries was a heavy dependence on imports a conscious policy goal. Rather, it was the consequence—the inevitable consequence—of industrialization in a situation of land scarcity.

In Worldwatch Paper 85, *The Changing World Food Prospect: The Nineties and Beyond*, published in 1988, I looked at the food situation in these three countries, noting the trends common to each of them. Seeing the remarkable consistency in their experiences and recognizing what this meant for China, I sent a copy of the study to Lin Zi Xin, head of the Institute for Science and Technology Information for China and a personal friend, to alert the leaders there to the potential for enormous growth in dependence on imported grain.

It was not until 1994, when working with my colleague Hal Kane on *Full House: Reassessing the Earth's Population Carrying Capacity*, that I turned again to China's food situation. Since the pace of industrialization there had accelerated, fueled by one of the world's highest savings rates and record foreign investment, I felt it useful to outline again what this would mean to China's food balance.

To do this, I wrote an article for *World Watch*, the Institute's magazine, looking at what rapid industrialization would mean for China's food balance and for the world if China followed the path of its three smaller neighbors, eventually importing most of its grain. Entitled "Who Will Feed China?" the article attracted more attention than anything I have ever written. In addition to appearing in the five language editions of our magazine—Japanese, Chinese, German, Italian, and English—it also appeared in many of the world's major newspapers, such as the *Washington Post, Los Angeles Times*, and *International Herald Tribune*. It was syndicated internationally by both the *Los Angeles Times* and the *New York Times*. Among the major news organizations covering the analysis were the Associated Press, Reuters, and the *Wall Street Journal*, including the Asian edition.

We released the article at a briefing in Washington, D.C., for the international press corps on Wednesday, August 24, 1994. The following Monday, the Ministry of Agriculture in Beijing held a press conference in which Deputy Minister Wan Baorui announced the Ministry's official disagreement with the analysis. He said that by 2025 they would nearly double their grain production, and thus would have no trouble satisfying their growing food needs. In fact, he said, the only grain they might import would be in the form of new seeds. Almost immediately, reporters were back to me asking for a reaction to the statement in Beijing.

Following my response, things quieted down until early in November when I was in Tokyo to receive the Blue Planet Prize, an annual award given by the Asahi Glass Foundation of Japan for environmental leadership. I was interviewed there by Reuters correspondent Eiichiro Tokumoto, who wanted me to elaborate on the China analysis. His story, carried on the Reuters world wire, was picked up in China.

Shortly thereafter an article appeared in the *China Daily* written by Hu An'gang, a research fellow with the Chinese Academy of Sciences in Beijing. Dismissing my analysis as unbelievable and unscientific, he likened it to a prediction by Secretary of State Dean Acheson some 45 years ago, who said that China would have great trouble feeding its 500 million people. Hu then pointed quite proudly, and correctly, to the dramatic gains made in grain production since the birth of modern China in 1949. He accepted my projection of the growth in grain demand in the decades ahead as incomes climbed, but rejected those for grain production. His main point was that China had an enormous remaining potential for expanding its food production that I was underestimating.

In early February 1995, I was in Oslo, Norway, to address an international conference of environment ministers hosted by Prime Minister Gro Harlem Brundtland. The theme of the conference was sustainable development. In my presentation, I sketched out a framework for sustainable development and illustrated some of the global dilemmas that lay ahead on the food front by outlining China's food prospect. I described China's likely emergence as a massive importer of food as "a wake-up call" that would force governments everywhere to address long-neglected issues, such as the need to stabilize population, to invest much more heavily in agriculture, and to redefine security in terms of food scarcity rather than military aggression.

Following the presentation, which was well received, I had to leave after the coffee break for the airport. Later I learned that when the session reconvened, the Chinese ambassador to Norway, Xie Zhenhua, asked for the floor even though he was not a scheduled speaker. He claimed that my analysis was off-base and misleading.

According to the *Times of India*, one of the papers covering my presentation and the ambassador's response, he said: "We are giving priority to agricultural productivity. Our family planning program has been very successful. Science and technology and economic growth will see us through." In concluding, he repeated my question "Who will feed China?" and solemnly replied that "the Chinese people will feed themselves." The following day, Ambassador Xie held a news conference, pointing out "unequivocally that China does not want to rely on others to feed its people and that it relies on itself to solve its own problems."

Although I was aware that the Chinese were sensitive to the notion that they might need to import large amounts of grain, I had not realized just how sensitive the issue is. All the leaders of China today are survivors of the massive famine that occurred in 1959–61 in the aftermath of the Great Leap Forward—a famine that claimed a staggering 30 million lives. If this many died, then as many as a couple hundred million more people could have been on the edge of starvation.

The national psyche of China clearly has been affected by this devastating famine. The prospect of depending on the outside world for a substantial share of the food supply is both psychologically difficult to accept and politically anathema. Those who are in leadership positions today are obviously reluctant to accept the notion that they are on a path heading toward heavy dependence on food from abroad. It is easy to sympathize with their concern.

Ironically, during the time when my indirect dialogue with Chinese officialdom was taking place, the food situation was tightening within China. A 60-percent rise in grain prices during 1994 led the government to buy from abroad a record 6 million tons of grain during one month at the end of that year, mostly from the United States, as it tried to check the price rise.

In late February and March of 1995, the tone of reports coming out of China began to change. On February 28, a Reuters story referred to the "sounding of alarm bells" by Communist Party chief and President Jiang Zemin and by Premier Li Peng about the state of China's agriculture. Premier Li talked about 1995 being "significant for the increase of grain and cotton output, and the task is a very hard one." President Jiang "warned that lagging agricultural growth could spawn problems that would threaten inflation, stability, and national economic development." He indicated that some developed coastal areas where industrialization was particularly rapid had suffered a precipitous drop in the amount of acreage under cultivation, saying that this is "a trend which must be reversed . . . this year."

At the National People's Congress meeting in mid-March, officials acknowledged, "China is facing a looming grain crisis, with a hike in imports the only apparent solution to the demands of a growing population on a shrinking farmland." Experts cited "a series of vicious circles that threatened to lock grain production into a downward spiral." Extensive consideration of the food issue at the Congress suggests that it is now becoming a matter of concern within official circles.

In early May 1995, I was invited to a dinner in Washington with Cheng Xu, Director of Science and Technology in China's Ministry of Agriculture. He shared with me a folder containing a photocopy of my article in the English edition of *World Watch* and a stack of responses to it, mostly in Chinese. Cheng said that the principal contribution of my article had been to focus the attention of China's leaders on agriculture, a sector they had neglected in their breakneck effort to industrialize. Whether China's political leaders are now ready to discuss publicly the dimensions of their likely future dependence on the outside world for food remains to be seen.

I wanted to write this book to document as carefully and clearly as possible what may lie ahead as this country of 1.2 billion people continues on its path of rapid industrialization. My aim is not to discourage China from moving in this direction, but rather to help us all to understand the consequences for China and the world of its doing so. The purpose of the book is not to blame China for the problems that are likely to arise from its projected emergence as a massive grain importer, but simply to recognize that this will force political leaders everywhere to recognize that the world is now on a demographic and economic path that is environmentally unsustainable.

Lester R. Brown

Who Will Feed China?

1

Overview:
The Wake-Up Call

We often hear that the entire world cannot reasonably aspire to the U.S. standard of living or that we cannot keep adding 90 million people a year indefinitely. Most people accept these propositions. Intuitively, they realize that there are constraints, that expanding human demand will eventually collide with the earth's natural limits.

Yet, little is said about what will actually limit the growth in human demands. Increasingly, it looks as though our ability to expand food production fast enough will be one of the earlier constraints to emerge. This is most immediately evident with oceanic fisheries, nearly all of which are being pushed to the limit and beyond by human demand. Water scarcity is now holding back growth in food production on every continent.

Agronomic limits on the capacity of available crop varieties to use additional fertilizer effectively are also slowing growth in food production.

Against this backdrop, China may soon emerge as an importer of massive quantities of grain—quantities so large that they could trigger unprecedented rises in world food prices. If it does, everyone will feel the effect, whether at supermarket checkout counters or in village markets. Price rises, already under way for seafood, will spread to rice, where production is constrained by the scarcity of water as well as land, and then to wheat and other food staples. For the first time in history, the environmental collision between expanding human demand for food and some of the earth's natural limits will have an economic effect that will be felt around the world.

It will be tempting to blame China for the likely rise in food prices, because its demand for food is exceeding the carrying capacity of its land and water resources, putting excessive demand on exportable supplies from countries that are living within their carrying capacities. But China is only one of scores of countries in this situation. It just happens to be the largest and, by an accident of history, the one that tips the world balance from surplus to scarcity.

Analysts of the world food supply/demand balance have recognized that the demand for food in China would climb dramatically as industrialization accelerated and incomes rose. They have also assumed that rapid growth in food production in China would continue indefinitely. But on this latter front, a closer look at what happens when a country is already densely populated before it industrializes leads to a very different conclusion. In this situation, rapid industrialization inevitably leads to a heavy loss of cropland, which can override any rises in land productivity and lead to an absolute decline in food production.

Historically, there appear to be only three other countries that were densely populated in agronomic terms before industrializing—Japan, South Korea, and Taiwan. The common experience of these three gives a sense of what to expect as industrialization proceeds in China. For instance, the conversion of grainland to other uses, combined with a decline in multiple cropping in these countries over the last few decades, has cost Japan 52 percent of its grain harvested area, South Korea 46 percent, and Taiwan 42 percent.[1]

As cropland losses accelerated, they soon exceeded rises in land productivity, leading to steady declines in output. In Japan, grain production has fallen 32 percent from its peak in 1960. For both South Korea and Taiwan, output has dropped 24 percent since 1977, the year when, by coincidence, production peaked in both countries. If China's rapid industrialization continues, it can expect a similar decline.[2]

While production was falling, rising affluence was driving up the overall demand for grain. As a result, by 1994, the three countries were collectively importing 71 percent of their grain. (See Figure 1–1.)[3]

Exactly the same forces are at work in China as its transformation from an agricultural to an industrial society progresses at a breakneck pace. Its 1990 area of grainland per person of 0.08 hectares is the same as that of Japan in 1950, making China one of the world's most densely populated countries in agronomic terms. If China is to avoid the decline in production that occurred in Japan, it must either be more effective in protecting its cropland (which will not be easy, given Japan's outstanding record) or it must raise grain yield

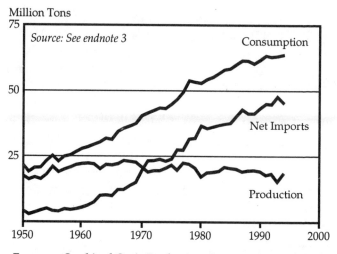

FIGURE 1-1. *Combined Grain Production, Consumption, and Trade for Japan, South Korea, and Taiwan, 1950-94*

per hectare faster during the next few decades than Japan has in the last few—an equally daunting task, considering the Japanese performance and the fact that China's current yields are already quite high by international standards.[4]

Building the thousands of factories, warehouses, and access roads that are an integral part of the industrialization process means sacrificing cropland. The modernization of transportation also takes land. Cars and trucks—with sales of 1.3 million in 1992 expected to approach 3 million a year by the decade's end—will claim a vast area of cropland for roads and parking lots. The combination of continually expanding population and a shrinking cropland base will further reduce the already small area of cropland per person.[5]

At issue is how much cropland will be lost and how

fast. Rapid industrialization is already taking a toll, as grain area has dropped from 90.8 million hectares in 1990 to an estimated 85.7 million in 1994. This annual drop of 1.26 million hectares, or 1.4 percent—remarkably similar to the loss rates of China's three smaller neighbors in their industrialization heyday—is likely to endure as long as rapid economic growth continues.[6]

China faces another threat to its food production that its three smaller neighbors did not. Along with the continuing disappearance of farmland, it is also confronted by an extensive diversion of irrigation water to nonfarm uses—an acute concern in a country where half the cropland is irrigated and nearly four fifths of the grain harvest comes from irrigated land. With large areas of north China now experiencing water deficits, existing demand is being met partly by depleting aquifers. Satisfying much of the growing urban and industrial demand for water in the arid northern half of the country will depend on diversions from irrigation.[7]

That China's grain production might fall in absolute terms comes as a surprise to many. This is not the result of agricultural failure but of industrial success. Indeed, China's record in agriculture is an exceptional one. Between 1950 and 1994, grain production increased nearly fourfold—a phenomenal achievement. After the agricultural reforms in 1978, output climbed in six years from scarcely 200 million tons to 300 million tons. With this surge, China moved ahead of the United States to become the world's leading grain producer. (See Figure 1–2.)[8]

Another way of evaluating China's agricultural record is to compare it with that of India, the world's second most populous country. Per capita grain production in China, which was already somewhat higher than in

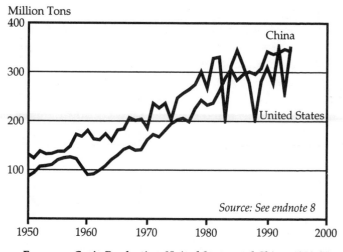

Million Tons

FIGURE 1-2. *Grain Production, United States and China, 1950-94*

India, climbed sharply after agricultural reforms were launched in 1978, opening an impressive margin over its Asian neighbor. (See Figure 1–3.)[9]

Between 1978 and 1984, China did what many analysts thought was impossible: In just six years, it raised annual grain production from roughly 200 kilograms per person to nearly 300 kilograms. At 200 kilograms, almost all grain is needed to maintain a minimal level of physical activity; an additional 100 kilograms a year opens the way for converting some grain into pork, poultry, and eggs. The immediate challenge facing China is not averting starvation, for it has established a wide margin between its current consumption level of 300 kilograms and the subsistence level. Rather, the challenge is to maintain price stability in the face of soaring demand for food driven by unprecedented advances in income.[10]

While China's food production capacity is starting to

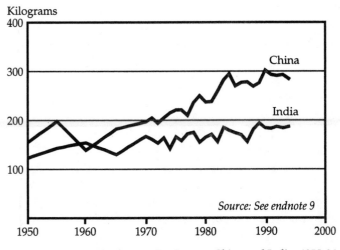

Kilograms

FIGURE 1-3. *Grain Production Per Person, China and India, 1950-94*

erode as a result of its breathtaking pace of industrialization, its demand for food is surging. The country is projected to add 490 million people between 1990 and 2030, swelling its population to 1.6 billion—the equivalent of adding four Japans. Because China's population is so large, even a slow rate of growth means huge absolute increases. Yet these increases are only the beginning of the story.[11]

Even as population expands, incomes are rising at an unprecedented rate. Economic growth of 13 percent in 1992 and again in 1993, of 11 percent in 1994, and of an estimated 10 percent in 1995 adds up to a phenomenal 56-percent expansion of the Chinese economy in just four years. Never before have incomes for so many people risen so quickly.[12]

This rapid economic expansion promises to push demand for food up at a record rate. When Western

Europe, North America, and Japan began establishing modern consumer economies after World War II, they were home to some 340 million, 190 million, and 100 million people, respectively. By contrast, China is entering the same stage with a population of 1.2 billion and an economy that is expanding twice as fast. If its rapid economic growth continues, China could within the next decade overtake the United States as the world's largest economy.[13]

Past experience has not prepared us well for assessing the scale of China's future food demand. Multiplying 1.2 billion times anything is a lot. Two more beers per person in China would take the entire Norwegian grain harvest. And if the Chinese were to consume seafood at the same rate as the Japanese do, China would need the annual world fish catch.

As incomes rise, one of the first things that low-income people do is diversify their diets, shifting from a monotonous fare in which a starchy staple, such as rice, supplies 70 percent or more of calories to one that includes meat, milk, and eggs. As consumption of pork, beef, poultry, eggs, milk, and other livestock products increases along with income, grain requirements rise rapidly.[14]

In neighboring Japan, the soaring demand for grain driven by prosperity combined with the heavy loss of cropland since mid-century to push dependence on grain imports to 72 percent of total grain consumption in 1994. These same forces are now at work in China. It is one thing for a nation of 120 million people to turn to the world market for most of its grain. But if a nation of 1.2 billion moves in this direction, it will quickly overwhelm the export capacity of the United States and other countries, driving food prices upward everywhere.[15]

The first signs of a growing imbalance between the demand and supply for grain in China became evident in early 1994. In February, grain prices in China's 35 major cities had jumped 41 percent over the same month in 1993. In March, driven by panic buying and hoarding, the rise continued unabated. In response, the government released 2.5 million tons of grain from stocks to check the runaway increase in prices. This calmed food markets, but only temporarily. By October, grain prices were 60 percent higher than a year earlier. More grain reserves were released, and the government banned trading in rice futures on the Shanghai Commodity Exchange. Speculators were driving futures prices upward, leading to panic among urban consumers. The 1994 inflation rate of 24 percent—the worst since modern China was created in 1949—was largely the result of rising food prices.[16]

Resisting the import of grain throughout most of 1994, Beijing let prices rise as much as possible to encourage farmers to stay on the land. In recent years an estimated 120 million people, mostly from the interior provinces, have left the land and moved to cities in search of high-paying jobs. This rootless, floating population, roughly the size of Japan's, wants to be part of the economic revolution. As a potential source of political instability, these migrants are a matter of deep concern in Beijing. The government is trying to maintain a delicate balance, letting the price of grain rise enough to keep farmers on the land but not so much that it creates urban unrest that could lead to political upheaval.[17]

Leaders in Beijing are also trying to deal with massive unemployment and underemployment, with much of the latter masked by villagers eking out a meager existence on tiny plots of marginal land. Creating enough jobs to employ productively an estimated 800 million workers depends on maintaining double-digit or near double-digit rates of economic growth. The government opened the country up to foreign investment in part because it was the only way to get the capital and technology needed to achieve this vital goal.[18]

If China holds together as a country and if its rapid modernization continues, it will almost certainly follow the pattern of Japan, South Korea, and Taiwan, importing more and more grain. Its import needs may soon far exceed the exportable supply of grain at recent prices, converting the world grain economy from a buyer's market to a seller's market. (See Chapter 7.) Instead of exporters competing for markets that never seem large enough, which has been the case for most of the last half-century, importers will be fighting for supplies of grain that never seem adequate.[19]

In an integrated world economy, China's rising food prices will become the world's rising food prices. China's land scarcity will become everyone's land scarcity. And water scarcity in China will affect the entire world.

In short, China's emergence as a massive grain importer will be the wake-up call that will signal trouble in the relationship between ourselves, now numbering 5.7 billion, and the natural systems and resources on which we depend. It may well force a redefinition of security, a recognition that food scarcity and the associated economic instability are far greater threats to security than military aggression is. The chapters that follow analyze this transformation, explaining why and how it is likely to come about.[20]

I

China: Taking Inventory

2

Another Half-Billion

As Chinese leaders analyzed future population, land, and water trends some 20 years ago, they realized that they had to choose between the reproductive rights of the current generation and the survival rights of the next generation. What separates the government in Beijing from those in many other countries is that it is desperately trying to protect the options of the next generation, politically difficult though that may be. This farsightedness and the political courage of the government of China deserve recognition.

In 1982, China's population reached 1 billion, making it the first member of an exclusive club. By 2017, its population is projected to reach 1.5 billion—equal to the world's entire population in 1900. Its demographic growth is then expected to slow and its population to peak at 1.66 billion in 2045, after which it should start to decline slowly. (See Figure 2–1.)[1]

Looked at in terms of the last four decades and the next four, the magnitude of China's population growth becomes clear. From 1950 to 1990, China added 571 million people. From 1990 to 2030, it is projected to add 490 million more. This anticipated addition reflects an impressive slowing in the rate of population growth, but it is still nearly a half-billion people. Stated otherwise, during the next four decades China will be adding an average of roughly 12 million people to the world annually.[2]

Many people think of Asia and Europe as having similar population densities, but in reality Asia has many more people per hectare of grainland than Europe does. The grainland per person in China today is roughly half that in France and it is inherently less fertile. The other

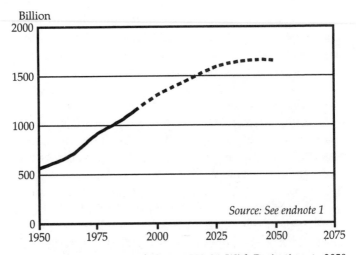

FIGURE 2-1. *Population of China, 1950-94, With Projections to 2050*

difference is that Europe's population has stopped growing. Now that the growth in grain use per person that accompanied rising incomes has also come to an end, the demands made by the region's population on its land and water resources have stabilized. Europe is living well within its food production capability or, in ecological terms, its food carrying capacity. Indeed, it is producing a modest grain surplus.[3]

In geographic area, China and the United States are essentially the same size. The big difference is that the western half of China is largely inhospitable to human habitation. Lacking basic life-supporting soil and water resources, it contains only a small percentage of the country's people.

For China, where the opportunity for bringing new land under cultivation is limited by water scarcity, population growth has a double-edged effect. First, it is shrinking the cropland area per person, as a fixed area is divided among an ever larger number of people. At the same time, the new citizens bring demands for living space, which in turn generates pressure to convert cropland to residential purposes. Simply housing an additional 490 million people in the next 40 years will require an enormous area, some of which will be cropland.[4]

There is a certain fascination with the demographic trends and issues of China partly because of its sheer size. In addition, the chaos of the Great Leap Forward and the arithmetic of the resulting famine that was long concealed from the outside world have intensified interest in China's demographic history.

In the late fifties, during the Great Leap Forward, millions of farmers were diverted to large construction projects, including roads, huge earthen dams, and backyard steel furnaces. This movement of labor from agriculture led to massive food shortages. Official records now show that 30 million Chinese starved to death during 1959–61. The demographic effect of the famine, however, extended far beyond these deaths.[5]

In populations that are near starvation, the frequency of intercourse decreases, sharply reducing the number of possible pregnancies. Beyond this, severely malnourished women cease ovulating, thus further reducing the number of pregnancies. Women who do conceive when they are severely malnourished often miscarry.[6]

During the heart of the famine in 1960, the number of deaths in China actually exceeded the number of births. The birth rate fell to 21 per 1,000 population while the death rate climbed to 25 per 1,000, leading to a decline in China's population for one year. (See Figure 2–2.)[7]

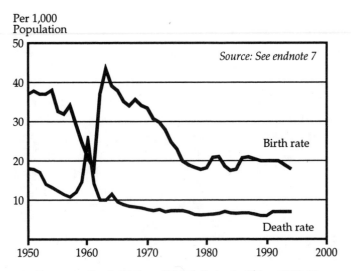

FIGURE 2-2. *Crude Birth and Death Rates in China, 1950-94*

The drop was so steep that it markedly affected the number of people added to world population. (See Figure 2–3.) Between 1950 and 1957, that number was increasing steadily, climbing from 39 million to 57 million. It then began to drop, reaching a low of 41 million in 1960. As China recovered from the famine, the annual addition climbed sharply to 70 million in 1962, recovering the trend that had existed before the famine.[8]

The Great Famine of 1959–61 left an indelible imprint on China's national psyche. John Bermingham, president of the Colorado Population Coalition, observes that "just as an American generation was seared by the Great Depression and a German generation by runaway inflation, the Chinese have had a generation seared by famine." These analogies help us understand the effect of the Chinese famine, but the latter was more

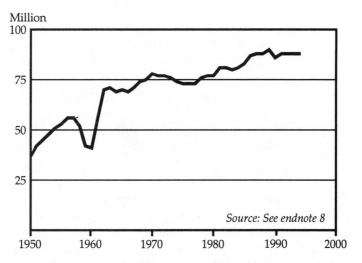

FIGURE 2-3. *Annual Addition to World Population, 1950-94*

traumatic simply because it was life-threatening for such a vast number of people.[9]

Like many governments, China was slow to recognize the population threat. Socialist ideology made it easy to dismiss the problem. As demographer Michael Teitelbaum notes, "For Marx, the fact that people were producers as well as consumers meant that the resource limits emphasized by the classical economists could arise under capitalism, but not under socialism."[10]

A quarter-century after the communist takeover in 1949, the government began to recognize population growth as a matter of concern. This occurred as part of the post-Mao reassessment in which projections of China's population growth were made. Much to the dismay of officials, based on an assumption of two children per couple, China could expect to add the better part of 1 billion people, adding the equivalent of an India to its existing population.[11]

As analysts attempted to calculate the consequences of population increase for cropland and water supply per person, and the availability of capital for creation of jobs, they realized that this was not a viable proposition. They were then confronted with one of the most politically difficult challenges for any government: the need to shift to a one-child family.

China faces some unique and difficult demographic issues. In contrast to Europe, there is no meaningful emigration safety valve. For countries such as England, Ireland, Germany, Italy, and Spain, mounting population pressure during their early development translated into a steady flow of migrants, many of them bound for the New World. Today, there are more people of Irish extraction in the United States than in Ireland. Similarly, there are more Spanish descendants in Latin

America than in Spain. Indeed, the New World is populated largely with the demographic overflow of Europe.[12]

The mounting population pressures in China today are occurring in a world far different from that of a century or two ago. No sparsely settled, habitable areas still exist. No country or group of countries wants to entertain the idea of absorbing 12 million Chinese each year.

Trying to put the brakes on population growth in China has not been easy. The government in Beijing, like those in many other developing countries, waited too long before implementing a meaningful effort to reduce family size. Faced with a tradeoff between smaller families in the present or deteriorating living conditions in the future as population pressures mounted, Chinese leaders opted in 1979 for the one couple/one child policy.

This policy, which explicitly reflects the interests of future generations, has run into heavy resistance. One source of difficulty has been a strong preference for male children, a desire so powerful, particularly in rural areas, that is has led to widespread female infanticide. In each annual cohort, males outnumber females until age 64; thereafter, females outnumber males. The conflict between local officials trying to implement this policy and couples intent on having more than one child has led, not surprisingly, to charges of coercion. It illustrates all too well the political conflicts that can develop within a society that is overrunning its human carrying capacity.[13]

Implementing the one-child-family policy has become more difficult in recent years. Job seekers migrating from countryside to city can more easily evade official monitoring of family size. Some families are becoming so affluent that they can readily pay the stiff penalty for having additional children. Moving quickly from a situation of rapid population growth to one of population stability has proved to be politically challenging to say the least.[14]

Nevertheless, it is possible to consider a scenario that would stabilize population size in China well below the 1.66 billion peak projected for 2045. The 1990 population pyramid, which gives the size of various age groups in the population, shows two age groups that are unusually small. (See Figure 2–4.) The first group, those who are 30–34 years of age, was reduced by the famine of 1959–61. The second, smaller group—those 10–14 years of age—shows the effect of family planning programs adopted in the mid-seventies and the echo effect of the smaller numbers born during the famine reaching reproductive age. If China can sustain its one-child-family program when the people born between 1975 and 1986 reach childbearing age, its population size could stabilize sooner rather than later, and far short of the projected 1.66 billion.[15]

In 1993, the Population Reference Bureau (PRB) pointed out that China had succeeded in lowering its fertility below replacement level—that is, the total number of children per woman was 2 or fewer. This was achieved within two decades of launching family planning programs.[16]

Carl Haub, senior demographer at PRB, noted that the birth rate was 21.1 per 1,000 population in 1990 and that it dropped to 19.7 in 1991 and to 18.2 in 1992. Since then the decline has continued, reaching 17.7 in 1994. This reduced China's population growth rate to 1.1 percent, roughly the same as that of the United States. The drop in China's population growth rate from 2.7 percent in 1970 to 1.1 percent in 1994 has

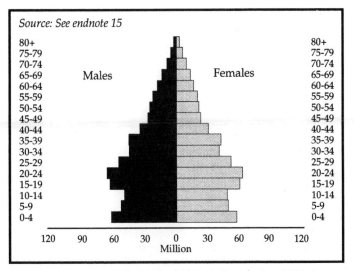

FIGURE 2-4. *Age Pyramid of China's Population, 1990*

played a key role in lowering the global population growth rate during that time from 2.1 percent to 1.6 percent.[17]

Recognizing the urgency of stabilizing population size, China's President and Communist Party chief, Jiang Zemin, renewed the call for one couple/one child in March 1995. Concern with the environmental, economic, and social effects of continuing population growth in China runs deep. President Jiang pointed out that "the rapid increase in a big population base has a direct bearing on the problems of food, of jobs, of education, of resource destruction, of environmental protection, and an imbalanced ecology." Increasingly, Chinese leaders are becoming aware of the environmental consequences of the combined effect of population growth and rising affluence that their country is experiencing.[18]

3

Moving Up the Food Chain

For those who live on the lower rungs of the global economic ladder, an increase in consumption of livestock products is a key measure of progress. When a Chinese villager in the interior of the country was asked in 1993 by a reporter if living conditions were improving, he responded, "Overall, life has gotten much better. My family eats meat maybe four or five times a week now. Ten years ago, we never had meat."[1]

China's 1.2 billion people are moving up the food chain at a remarkable rate. As noted in Chapter 1, China has experienced a phenomenal 56-percent expansion of its economy in just four years. With population growing at scarcely 1 percent per year, income per person has risen by half within this period.[2]

Never in history have so many people moved up the

food chain so fast. As incomes rise, one of the first things that low-income people do with their money is diversify their diets, shifting from a daily menu in which a starchy staple such as rice supplies 70 percent or more of calories to a more diverse fare including meat—pork, poultry, beef, and mutton—and eggs, milk, butter, cheese, yogurt, and ice cream. This pattern of dietary change as incomes climb is common to all societies with the exception of those where religious restrictions apply, such as on beef in Hindu societies or pork in Muslim countries.[3]

The amount of livestock products consumed per person varies greatly among countries. For beef, for example, the average American eats 42 kilograms per year compared with 1 kilogram in China. For pork, consumption levels are much closer: 28 kilograms versus 21 kilograms. Neither society consumes much mutton, on average in both about 1 kilogram per person a year. (See Table 3–1.)[4]

Countries living high on the food chain, such as the

TABLE 3-1. *Annual Per Capita Grain Use and Consumption of Livestock Products in Selected Countries, 1990*

Country	Grain Use[1]	Consumption					
		Beef	Pork	Poultry	Mutton	Milk[2]	Eggs
		(kilograms)					
United States	800	42	28	44	1	271	16
Italy	400	16	20	19	1	182	12
China	300	1	21	3	1	4	7
India	200	—	0.4	0.4	0.2	31	13

[1]Rounded to nearest 100 kilograms since the countries are selected to show the range of consumption categories. [2]Total consumption, including that used to produce butter, cheese, yogurt, and ice cream.

SOURCE: See endnote 4.

United States and Canada, use some 800 kilograms of grain per person a year as food. Some of it is consumed directly as bread, breakfast cereals, and pastries, but most of it is eaten indirectly in the form of livestock products. In affluent societies, a small fraction of grain is also consumed in the form of grain-based beverages, such as beer, bourbon, scotch, and vodka. The total use of grain per person among the highest and lowest ranges from roughly 800 kilograms in the United States to 200 kilograms in India, a ratio of 4 to 1. In contrast to energy, where the variations among countries can easily reach 30 to 1, consumption of food per person among countries varies in a much more narrow range.[5]

For Americans, the vast rangelands of the Great Plains sustain huge herds of beef cattle that are then finished in feedlots. Although large amounts of grain are fed to cattle in feedlots, the major share of U.S. beef production comes from grass. China, however, does not have vast rangelands. If it wants to produce more beef, it will have to do so largely in the feedlot.[6]

The growth in meat production and consumption in China since the economic reforms of 1978 is one of the best measures of the country's economic transformation. In 1977, consumption totalled 7.7 million tons. By 1994, this had climbed to 40 million tons, a fivefold increase in 16 years. Meat consumption per person climbed from 8 kilograms in 1977 to more than 32 kilograms in 1994. The gap in red meat consumption between China and industrial countries is narrowing. Because of China's high pork consumption per person, its total red meat consumption is now substantially larger than that of the United States.[7]

Indeed, pork dominates China's meat consumption, accounting for some three fourths of the total in 1994. (See Figure 3–1.) Lacking the rangelands to support

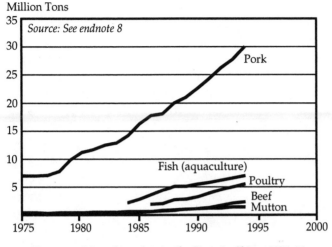

FIGURE 3-1. *Meat Consumption by Type in China, 1975-94*

large herds of cattle or flocks of sheep, China has relied heavily on pigs for meat. The typical village family often has one pig, which provides a way of both disposing of table waste and providing fertilizer.[8]

The consumption of poultry, although it started from a much smaller base, is now expanding even more rapidly. Between 1986 and 1994, output increased from 1.9 million tons to 5.5 million tons, tripling in eight years.[9]

As China looks to consuming more meat in the future, it must also look for more grain. More meat means more grain—two kilograms of additional grain for each kilogram of poultry, four for pork, and seven for each kilogram of beef added in the feedlot. As the Chinese get richer, they will eat more meat, milk, and eggs, but if the supply of grain does not expand apace with their appetites, food prices will soar.[10]

With pork consumption already near western levels,

the government is emphasizing the expansion of poultry, eggs, and beef production. Although pork consumption per person in China is approaching that of the United States, the intake of beef, poultry, and dairy products is still minuscule by comparison. Poultry was once a rare luxury in China, and the average person still eats only one tenth as much chicken as an American. But the appetite for poultry is growing fast. Ironically, that change has been spurred by a government policy that encourages production of chickens because they convert grain into meat more efficiently than pigs or cattle do. During the nineties, poultry consumption, starting from a small base, is expanding at double-digit rates.[11]

The same thing is beginning to happen with eggs. The official goal for egg consumption has been set at 200 per person by the year 2000—double the 100 of 1990 and close to the 235 consumed per year by the average American. With the population expected to reach 1.3 billion people by then, annual egg consumption will rise to 260 billion. If eggs laid per hen in the year 2000 reach 200 per year (U.S. hens average 250 a year), China will need a flock of 1.3 billion hens to satisfy this need. Yet reaching this goal will take an additional 24 million tons of grain, an amount equal to the total exports of Canada, the world's second-ranking exporter.[12]

China is emphasizing the increased production of beef, but it is also starting to import some. In 1993, the nation brought in 2,100 tons of beef, much of it high-quality beef from Australia. This is purchased especially for use in hotels and restaurants to satisfy the expectations of visitors from abroad, including a growing number of tourists and business community representatives. Whether China gets beef by producing or importing it, supplies will almost certainly have to come from feed-lots, simply because nearly all the world's rangelands are now being grazed at or beyond capacity.[13]

Consumption of dairy products in China is among the lowest anywhere, partly because it has one of the least lactose-tolerant populations of any country. (A Chinese restaurant menu several pages long typically will not offer a single dish that incorporates dairy products.) Consumption of milk in China totals 4 kilograms per person a year, compared with 271 kilograms in the United States. In India, where a huge herd of cattle supplies milk in addition to draft power, consumption per person is 31 kilograms. Despite widespread lactose intolerance, China's consumption of dairy products, particularly ice cream, is likely to rise steadily in the years ahead, following the trend of its affluent East Asian neighbors.[14]

For China, it would be tempting to turn to the oceans for its animal protein as population pressure on the land intensifies, much as Japan did. As land became scarce there beginning a century ago, that country began relying on the oceans for its animal protein. The result was the fish and rice diet that now characterizes Japanese cuisine.

Today, Japanese consumption of seafood is some 81 kilograms per person a year, one of the highest in the world. Japan's annual total take from the oceans is roughly 10 million tons. If China, with a population that is 10 times larger, were to turn to the oceans for a similar dependence on seafood for animal protein, it would need 100 million tons of seafood—an amount that matches the world fish harvest of 101 million tons in 1994.[15]

In recent years, as fleets of other seafood-hungry

countries have joined Japan in the aggressive pursuit of fish, oceanic fisheries have been pushed to their biological limits. According to the U.N. Food and Agriculture Organization, all 17 of the world's major fisheries are being fished at or beyond capacity. Nine are in a state of decline. So the Japanese option has been eliminated for any major newcomers.[16]

Future growth in demand for fish in China will have to be satisfied largely by fish farming. Faced with the need to cultivate its own fish supply, the nation has been producing some 6 million tons of fish (mostly carp) a year. This, in turn, increases the demand for grain by roughly 2 tons for each ton of fish produced, putting yet another demand on the country's shrinking grain fields. Rising grain prices, combined with the need to use scarce land and water for fish farming, will constrain the growth in fish consumption.[17]

Since the agricultural reforms of 1978, China's use of feedgrains has increased steadily, approaching 80 million tons in 1994—some 23 percent of total grain consumption. (See Figure 3–2.) Its total use of feedgrains now ranks second only to that of the United States. If incomes continue to rise in the years ahead, then feedgrain use will also keep rising, absorbing an ever larger share of the world's total grain supply.[18]

And the good life for the newly affluent Chinese does not stop with meat and fish. They are also acquiring a great enthusiasm for beer. In 1981, beer production totalled roughly 1 billion liters. By 1994, it had climbed to some 13 billion liters, or 11 liters per person. Over the next four years, output is projected to reach 22 million liters.[19]

To raise beer consumption for each adult by just one bottle takes an additional 370,000 tons of grain. Three

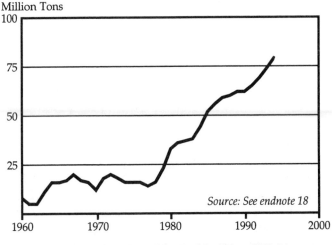

Million Tons

FIGURE 3-2. *Grain Used for Feed in China, 1960-94*

additional bottles per person would take the equivalent of Norway's entire grain harvest. This trend in China appears to be closely tracking that in Japan, where beer consumption per person is now among the highest in the world. Already China has overtaken Germany in total beer consumption, leaving it second only to the United States.[20]

In addition to the shift to more animal protein as incomes rise, people on the low rungs of the global economic ladder also want to use more vegetable oil. In 1994, China's consumption of vegetable oil per person was 6 kilograms per year. In Japan, it is 12 kilograms per year, and in the United States, 23. If China were to double its use to 12 kilograms, it would require more than 7 million additional tons of vegetable oil. Satisfying this new demand from imports would require world vegetable oil exports, currently at 23 million tons (of which close to half is soybean oil), to expand by nearly one third. The United States, the world's leading producer and exporter of soybean oil, exports 7 million tons of soybean oil a year.[21]

The U.S. agricultural attaché in Beijing estimates that in the marketing year beginning in 1995, China will be consuming 9.3 million tons of soybean oil and importing 3.5 million tons. Between 1984 and 1995, China will have gone from being self-sufficient in vegetable oil to importing 38 percent of its total consumption, with the growth in imports concentrated in the last few years. (See Figure 3–3.)[22]

Another commodity whose consumption climbs rapidly with income is sugar. China's sugar consumption of 6 kilograms per person annually is among the lowest in the world. The figure in India, which has a sweet tooth

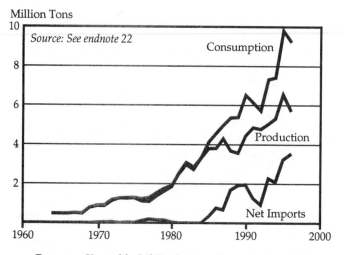

FIGURE 3-3. *Vegetable Oil Production, Consumption, and Trade in China, 1964-96*

compared with other low-income countries, is 20 kilograms per year, much of it consumed in tea. As of 1994, China used 7.2 million tons of sugar and produced 6 million tons, leaving it dependent on imports for 1.2 million tons—17 percent of its consumption. As sugar intake rises, China's claims on the world sugar harvest will increase accordingly.[23]

With the cropland base shrinking and with water shortages spreading, much if not all the growth in demand for food in China translates directly into imports. As the demand for grain, vegetable oils, and sugar continues to rise, and as production increases little or even declines, the difference between the two must come from imports. Given the scale of China's projected needs, this will put pressure on world supplies, affecting food prices everywhere.

4

The Shrinking Cropland Base

In 1984, the government of China issued a directive restricting use of the traditional burial mounds because they were consuming too much scarce land. Instead, Beijing urged cremation. This competition for land between the living and the dead poignantly underlines the extent of China's land scarcity.[1]

China is a large country, but only one tenth of its land is cultivable. Simply stated, much of this vast country is desert and mountain. And most of the cultivable land is in a 1,500-kilometer strip along the eastern and southern coasts.[2]

Not only is the cropland area per person one of the world's smallest, it is shrinking rapidly. Between 1990 and 1994, the grain area harvested dropped from 90.8 million hectares to 85.7 million. This decline of 5.6 percent in four years, combined with a population growth of 59 million (4.9 percent), reduced the grain harvested area per person by a striking 10.5 percent.[3]

The loss of cropland has long been a matter of concern to China's leaders, but with the heavy losses of the early nineties, the issue acquired a new sense of urgency. In early 1995, Zou Yuchuan, director of China's State Land Administration, said, "We have been racking our brains to find ways of protecting our farmland. But we are facing difficulties formulating policies to deal with the problem—the situation changes all the time." In referring to the heavy loss of cropland between 1990 and 1994, the *Economic Information Daily* of China said that "a land crisis is approaching. . . . If farmland loss continues at the present rate, China will suffer a serious problem of lack of food by the beginning of the twenty-first century."[4]

As indicated in Chapter 2, Asia is much more densely populated than Europe. Europe, whose population has stabilized, has a larger grain harvested area per person than does China, where population is projected to expand by some 40 percent before stabilizing. Beyond this, heavy industrialization has been completed in Europe, thus limiting future nonfarm demands on cropland. In China, most of the industrialization needed to create a modern consumer economy still lies ahead.

Understanding what lies in store for China depends on some knowledge of how rapid industrialization has affected the cropland area in countries that were already densely populated before serious industrialization began. There are, as noted earlier, only three countries that fit into this category: Japan, South Korea, and Taiwan.

The shrinkage in the grainland area in the three countries is remarkably similar. After peaking in 1955, Japan's grainland area shrunk by 52 percent over the roughly four decades to 1994, or some 1.4 percent a year. For South Korea, the area has dropped 46 percent since peaking in 1965, an annual decline of 1.2 percent. The trend for Taiwan is similar—a loss of 42 percent from 1962 to 1994, or 1.2 percent a year. (See Figure 4–1.) For the three countries combined, the grainland area peaked in 1956 at 7.9 million hectares; by 1993, it had declined to 4.1 million hectares. This drop of 48 percent over 37 years means the grainland area shrank by an average of 1.2 percent a year.[5]

The remarkable consistency in the effect of industrialization on the cropland base suggests a certain inevitability. And it indicates how difficult, if not impossible, it will be for China to avoid a similar loss of cropland.

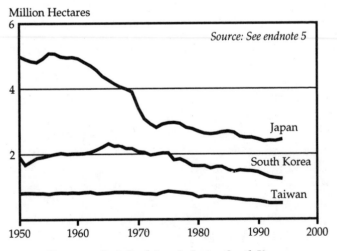

FIGURE 4-1. *Grainland Area in Japan, South Korea, and Taiwan, 1950-94*

In countries that are densely populated before industrialization begins, at least four distinct trends affect the
grain harvested area. One is the conversion of cropland
to nonfarm uses, including the construction of factories,
housing, and roads. Another is the abandonment of
cropland that is marginal either because its fertility is
low or because plots are so small. A third is a decline in
multiple cropping. The latter two trends are accelerated
by rising wages in the nonfarm sector. And finally, as
rising incomes drive up the demand for and prices of
fresh fruits and vegetables, farmers shift land from grain
to more profitable crops.

Creating jobs in industry, like creating jobs in agriculture, requires land. Today China's labor force totals
nearly 800 million, most of whom work in agriculture.
Shifting 100 million workers from the farm labor force
to the industrial sector, broadly defined, and assuming
100 employees per industrial establishment (about par
for China's private sector), means building 1 million
factories. Each factory needs a warehouse to store the
raw materials used in manufacturing and to store finished products until they are shipped. Each factory also
needs an access road. Factories have to be built where
the people are, and the people are concentrated where
the cropland is. Thus, modernization of the Chinese
economy, as with Japan, South Korea, and Taiwan
before it, means sacrificing cropland.

Residential demands are also claiming cropland. The
490 million people to be added to China's population
between 1990 and 2030 need to be housed. If it is assumed that each family consists of five individuals—a
married couple, one child, and one set of in-laws—the
additional people will require 98 million more housing
units. Whether this need is satisfied with apartments or

freestanding homes, it will consume a vast area of land, much of it cropland.[6]

Another consequence of rising affluence is an increase in the living space per person. In many cases, villagers are expanding their homes, adding a room or two. Others are simply building new, much larger homes. To illustrate, the floor space per person in Japan expanded from 20.5 square meters in 1970 to 28.6 square meters in 1990, an increase of more than one third in 17 years. Given the nationwide shortage of housing and the extent of crowding, this trend can be expected to continue as long as incomes rise.[7]

Automobiles, too, are "consuming" cropland. In an industrial policy announced in July 1994, Beijing indicated that automobiles are to become one of the four growth industries of the next two decades—along with telecommunications, computers, and petrochemicals. Ironically, even as the Ministry of Agriculture is calling for new measures to protect cropland, the Ministry for Machinery Building is pressing for a massive expansion of the automobile fleet and is planning to provide incentives to encourage people to trade their trusty bicycles for cars.[8]

Annual sales of cars, vans, trucks, and buses, which totalled 1.2 million in 1992, are expected to approach 3 million by decade's end. By 2010, Ministry projections show production of automobiles alone above 3.5 million per year, with two thirds of the new vehicles being sold to private owners.[9]

Meanwhile, the fleet of cars, which has increased from 1.15 million in 1990 to 1.85 million in 1994, is projected to reach 22 million by 2010. (See Figure 4–2.) A fleet of this size will require millions of hectares of land for a network of roads, highways, service stations,

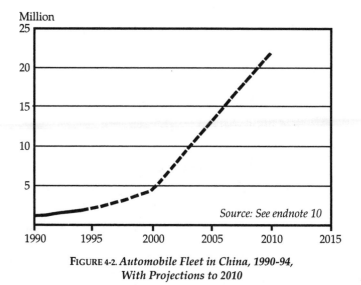

FIGURE 4-2. *Automobile Fleet in China, 1990-94,*
With Projections to 2010

and parking lots. Again, as with factories, these will have to be built where the people are.[10]

The Chinese government is now launching a massive highway construction program. Among other things, this includes four trunk routes that span the country in different directions. The longest one will link industrial northeastern Heilongjiang province with the southern island of Hainan. The second will go from Beijing to southern Guangdong province, ending up near Hong Kong. A third will link Jiangsu province on the east coast with Xinjiang in the far northwest. The fourth will link Shanghai, the city that is emerging as the country's major commercial center and port, with Sichuan province in the southwest.[11]

Builders completed 10,000 miles (16,129 kilometers) of new highway in 1994, of the 11,000 miles planned for that year. To reach its goal of building 1,800 miles of expressways by the end of this decade, the government is encouraging private investment in the construction of toll roads. The first such investment by a Hong Kong businessman led to the construction of a 77-mile superhighway linking Hong Kong and Guangdong. Projected to be self-financing, this $13-per-vehicle toll road has consumed a broad swath of cropland, including some of China's most productive riceland.[12]

The bottom line is that there is now an army of bulldozers cutting their way through the Chinese countryside building highways across rice paddies and wheat fields, leaving the nation with ever less cropland to satisfy its rising demand for food.

In addition to factories, housing, and roads, farmland is also being claimed by shopping centers, tennis courts, golf courses, and private villas. In rapidly industrializing Guangdong province, an estimated 40 golf courses have been built in the newly affluent Pearl River Delta region alone. Concern about this wholesale loss of cropland has led the Guangdong Land Bureau to cancel the construction of all golf courses planned but not yet completed.[13]

As noted earlier, the rapid rise in nonfarm wages now so evident in China typically leads to the abandonment of marginal cropland. Farmers with small holdings of the least productive land find they cannot participate in the rise in living standards unless they head for the cities. In effect, rising incomes in the society make the cultivation of marginal land unprofitable. They also lead to a decline in multiple cropping. Rising grain prices can slow this process, but even in Japan, where rice is supported at six times the world market price, these trends continue with little abatement.[14]

One of the keys to boosting food production in Asia,

particularly during this century's third quarter, was a rise in multiple cropping—an increase in the number of crops produced per hectare of cropland per year. By 1964, for example, Taiwanese farmers had boosted the number of crops per hectare to a national average of 1.9. In effect, they were harvesting on average almost two crops for their entire cropland area. In Japan, which has a shorter growing season since it is further north, the multiple cropping index reached 1.33 in 1960. South Korea's peaked in 1963 at 1.45. Interestingly, in each of these three countries, the multiple cropping index appears to have peaked during the early to mid-sixties and then begun to decline.[15]

The index drops as industrialization progresses and wages rise. Multiple cropping is a labor-intensive activity during certain periods, dependent on quickly harvesting a crop once it is mature and preparing the seedbed and planting the next one. As nonagricultural wages rise, it becomes more difficult for farms to compete for the labor needed during the peak periods of field activity, and multiple cropping declines. In Taiwan, it dropped to 1.23 in 1993. In South Korea, the figure fell to 1.08 in 1992. And in Japan, by 1988 it was 1.03, scarcely one crop per hectare. At this point, the Japanese government apparently discontinued this indicator because the remaining amount of multiple cropping was negligible. (See Figure 4–3.)[16]

For China, the multiple cropping index reached 1.5— that is, an average of 1.5 crops per hectare—for the first time in 1978, the first year of the reforms. By 1992, it had edged up to 1.56 percent. China may now be at the point where rising wages will lead to a gradual decline in multiple cropping, as it did in the other Asian nations three decades earlier. Indeed, the 5.6-percent decline in

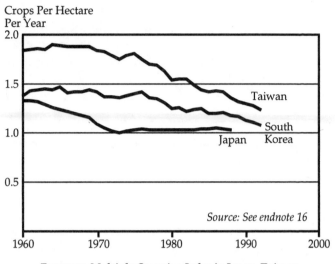

FIGURE 4-3. *Multiple Cropping Index in Japan, Taiwan, and South Korea, 1960-92*

grain harvested area between 1990 and 1994 suggests that the multiple cropping index may be peaking, setting the stage for a gradual, long-term decline.[17]

Closely associated with rising affluence is a growing demand for fresh fruit and vegetables. As a result, the area in high-value vegetable crops, which are more profitable than cereals, increases rapidly. Between 1970 and 1994, the area in vegetables more than tripled, from 2.7 million hectares to 8.7 million. (See Table 4–1.)[18]

If incomes continue to rise rapidly, the demand for fresh vegetables will also continue to rise as the Chinese consume more strawberries, asparagus, lettuce, and other highly valued garden crops. There is no reason to assume that the conversion of grainland to vegetable crops will not continue during the next few decades.

Nowhere have the forces just described been stronger

TABLE 4-1. *China: Area in Vegetables, 1970–94*

Year	Area
	(million hectares)
1970	2.7
1979	3.2
1981	3.4
1982	3.9
1983	4.1
1984	4.3
1985	4.7
1986	5.3
1987	5.6
1988	6.0
1989	6.3
1990	6.4
1991	6.5
1992	7.0
1993	7.9
1994	8.7

SOURCE: See endnote 18.

than in China's booming coastal provinces in the south. Here, the land where factories now stand was just a few years ago producing two or three crops of rice per year. Losing this means losing some of the world's most productive cropland.

If the nation continues on essentially the same industrial path as that followed by Japan, South Korea, and Taiwan, and if this reduction of grainland continues, China will have lost roughly half its grainland by 2030. If the population continues to grow as projected, adding 490 million people between 1990 and 2030, the grainland area per person will shrink from 0.08 hectares in 1990 to 0.03 hectares in 2030. (See Figure 4–4.)[19]

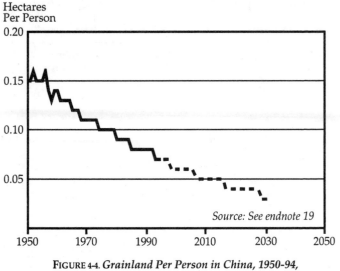

Hectares
Per Person

FIGURE 4-4. *Grainland Per Person in China, 1950-94,
With Projections to 2030*

Chinese political leaders are tempted to think that they can somehow avoid a massive loss of cropland to nonfarm uses. But if they look at the experience in Japan, they will see how difficult protecting cropland can be. Few governments have worked as strenuously at this as Japan did. Even within the city limits of Tokyo, tiny patches of land produce rice. Some 13,000 Japanese families farm land within that city. If this land were released for sale in the Tokyo real estate market, it might easily be worth 100 times its value as farmland. It is difficult to see how China, which lacks the social cohesion and discipline of Japan, could do any better. Even with this effort, Japan's harvested grainland area has shrunk by half during the last four decades.[20]

This seemingly inevitable loss of land is not so much a matter of land use policy as it is a function of population density relative to cropland. If China's leaders try hard, they can slow the loss of cropland, but they cannot prevent it. And this loss will increase pressure on cropland elsewhere in the world.

5

Spreading Water Scarcity

As recently as mid-century, water supplies in China were abundant, relative to demand. Surface and underground sources together were more than adequate to satisfy the needs of the country's 500 million people. Since then, however, the water supply/demand balance has deteriorated as water use has increased sixfold—driven by population growth, irrigation expansion, rising affluence, and industrialization. The water scarcity that now plagues much of China reflects the extent to which demand is outrunning the sustainable yield of rivers and aquifers.[1]

In late 1993, Minister of Water Resources Niu Mao Sheng observed that "in rural areas, over 82 million people find it difficult to procure water. In urban areas, the shortages are even worse. More than 300 Chinese cities are short of water and 100 of them are very short."[2]

In large areas of north China, demand is being met by depleting aquifers. In much of the country, future urban and industrial demand can be satisfied only by diverting water from irrigation—a worrying development in a country where half the cropland is irrigated.[3]

Water scarcity is one of the more difficult issues facing the government of China. As *New York Times* Beijing correspondent Patrick Tyler notes, "Any threat to China's ability to provide enough water for food production, job creation and overall economic growth is a threat to the established order."[4]

The area of irrigated land in China, totalling some 17 million hectares in 1950, reached 47 million hectares in the nineties—a near tripling. (See Figure 5–1.) This four-decade trend, however, masks two distinctive periods. During the first, from 1950 to 1977, the irrigated area increased by more than a million hectares annually. From 1977 to 1991, it increased by only 170,000 hectares a year. An analyst from the East-West Center in Hawaii argues that there has been very little real growth since the agricultural reforms of 1978.[5]

From mid-century through 1977, the growth in irrigated area greatly outstripped that of population, expanding the irrigated area per person by roughly two thirds, from 0.03 hectares in 1950 to 0.05 hectares in 1977. The launching of the economic reforms in 1978 brought this large investment in irrigation to an end. From 1977 onward, the irrigated area per person has declined, dropping roughly one fifth, or back to 0.04 hectares per person, by the early nineties.[6]

With 49 million hectares, China has more irrigated land than any other country. This compares with some

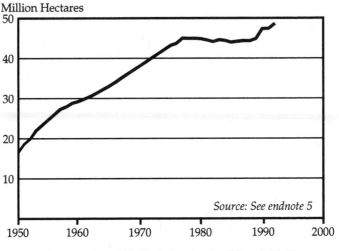

Million Hectares

FIGURE 5-1. *Land Under Irrigation in China, 1950-92*

46 million hectares in India and 20 million in the United States, the countries ranking second and third in irrigated area. Far more important to China than to those other nations, irrigation covers roughly half of the total cropland area and accounts for nearly four fifths of the all-important grain harvest.[7]

In the four-and-a-half decades since the Communist regime took over, the growth in irrigation has come from two sources. Until about 1970, most of it came from the development of surface water resources with dams, both large and small. The water was backed up and then diverted onto the land, usually through a gravity-fed system of canals. Since 1970, however, most of the additional irrigation has come from tapping underground water. Some 2 million wells now supply water for irrigation.[8]

Irrigation has led to much more intensive land use. In

some cases, land that once produced a single crop of rice now produces two or three crops a year. In other cases, wheat is grown as a winter crop and rice as a summer crop, both irrigated. The net effect was a substantial increase in the multiple cropping index from 1950 to 1980, as noted in Chapter 4—an increase in the harvested area without any appreciable gain in the actual cropland area. Now this era may be coming to an end.[9]

One of the best analyses of China's changing water balance appears in *China's Environmental Crisis* by Vaclav Smil, a China scholar at the University of Manitoba. Smil describes in dire terms the extent of falling water tables in northern China, noting that the province most affected by overpumping is Hebei, which surrounds both Beijing and Tianjin.[10]

Smil notes the distinction in northern China between the area of irrigated land and the land adequately irrigated. Some of the irrigated land receives only a fraction of the water needed to maximize yields. Hebei, Hunan, Shandong, and Shanxi are among the provinces where the application of irrigation water is far less than the ideal. Smil notes that "growing water claims of the plains, large cities and industrial areas will tend to lower even these modest irrigation rates."[11]

Water scarcity is most acute in Shanxi province, where one tenth of the province's peasants face chronic shortages of drinking water. Not only is the amount of irrigation water applied less than the optimum, but one quarter of the province's irrigated fields cannot be guaranteed water during the growing season.[12]

One of the cities in the 100 facing severe water shortages is Beijing, the capital. In the region immediately around the city, farmers in early 1994 were banned from the reservoirs from which they once drew their irrigation water. Burgeoning urban water demand now requires that all local water be available for use in Beijing.[13]

In the competition between the countryside and the city for water supplies, one that is now being waged at countless sites both in China and around the world, farmers invariably lose. They simply cannot afford to pay nearly as much as residential and industrial users for water. With 300 cities in China already short of water, it is likely that in the years ahead farmers will be excluded from water sources surrounding many other urban areas as water scarcity intensifies. More and more farmers will be forced to join those around Beijing who have reverted to less intensive rain-fed farming.[14]

As farmers turn increasingly to groundwater, the pumping of underground water in large areas of northern China is exceeding the recharge rate of aquifers. The result is falling water tables. In southwestern Shanxi province, for example, overpumping has dropped the water table by some 70 meters (or nearly 230 feet). As aquifers are depleted, land levels often fall. This process, known as subsidence, now affects an area in northern China the size of Hungary.[15]

Subsidence can affect aquifers' inherent long-term productivity. Aquifers have two functions: the storage and transport of water. Both are important and both may be diminished by excessive pumping. Unfortunately, the long-term geological effects of progressive aquifer depletion are poorly understood. Are the parts of the world now suffering from subsidence experiencing irreversible declines in the water storage and transport capacity of their aquifers?

In regions of China where pumping rates exceed aquifer recharge rates, the amount of water pumped will eventually be reduced. This, in turn, means either using water more efficiently, where that is possible, or shifting to less intensive cropping practices. In central and northern China, for example, where wheat and rice are double-cropped, this may mean replacing rice with a less water-demanding, lower-yielding staple crop, such as sorghum or millet. This shift to less intensive farming may arrest the fall in the water table, but it is not a welcome prospect in a country where the demand for grain is growing at a record rate.

Another form of damage from surface water irrigation is waterlogging and salinity. These are reducing productivity on an estimated 15 percent of China's irrigated land. When river water is diverted onto the land, part of it percolates downward. Without adequate drainage, the water table rises. When it reaches a few feet below the surface, deep-rooted crops suffer from waterlogging of their root zones. When it gets within inches of the surface, water evaporates through the soil into the atmosphere, leaving a thin layer of salt on the soil surface. Unless this rise in the water table can be reversed, by installing either more wells in the area or an underground drainage system, the accumulating salt eventually turns fertile land into wasteland, as it did in the early Middle East civilizations.[16]

Future water needs in China are expected to continue growing at a rapid pace. Each sector—agriculture, industry, and residential—will be demanding far more water a few decades from now than it is today. With food, for example, the combination of a population reaching 1.6 billion by 2030 and the continuing rise of individual consumption of livestock products could nearly double the demand for grain over current levels.[17]

Although historically the industrial sector has not accounted for a large share of water use in China, its demand for water has started to soar. With an industrial growth rate of more than 11 percent a year, industrial water needs could easily double within seven years. In Beijing, 23 percent of water withdrawals are for industry. In Tianjin, the figure is slightly lower, at 19 percent. One consequence of acute water scarcity in the northern part of China could be the gradual shift of more water-intensive industries to the water-rich southern provinces.[18]

Similar huge growth in water demand is in prospect in the residential sector. Today only a very small percentage of China's 1.2 billion people live in homes with indoor plumbing, but virtually everyone aspires to such a home and millions more are reaching this goal every year. Families living in traditional style, often sharing sanitary facilities, may use only 10 gallons of water per person daily. Those living in more modern apartment houses easily use two to three times this amount. As incomes rise, so does water consumption. As this transition to modern living progresses, residential water needs will also climb.[19]

Although there are no good data on the extent of water diversion from farming to the industrial and residential sectors, anecdotal evidence—such as the banning of farmers from reservoirs in the vicinity of Beijing—indicates it is growing rapidly.[20]

One reason for severe water scarcity in the northern half of China is the regional imbalance between the distribution of water and the distribution of cropland. More than four fifths of the surface water in China is found in the Yangtze and other river basins in the south. But this region has only 37 percent of the country's cropland. The large area north of the Yangtze, which has one fifth of the surface water, has nearly two thirds of the cropland.[21]

Concern over water scarcity is rising, particularly in the north. Water shortages in northern China have even raised questions about the suitability of Beijing as the capital. They have also renewed discussions of a 1,400-kilometer (860-mile) canal that would bring water from the south to the water-deficit north. Although the cost of building this enormous conduit—comparable to supplying Washington, D.C., with water from the Mississippi River—was initially estimated at $5 billion, the total could ultimately be several times larger. Among other things, such a canal will challenge engineers because it must cross 219 rivers and streams, including the Huang He (Yellow River), en route to Beijing.[22]

The water for this South-North Water Diversion Project would come from a reservoir on a tributary of the Yangtze River at Danjiangkou. If this reservoir is expanded and the water in it raised by 50 feet, it will have a drop of 300 feet to Beijing, making it possible for the water to move over the 1,400-kilometer aqueduct largely by gravity. As it approaches Beijing, the aqueduct is slated to divide, with one branch going to the capital and the other to Tianjin.[23]

This system of transferring water from one river basin to another would carry seven times as much water as is now used in New York City. This huge conduit would be designed to irrigate some of the land along its 1,400-kilometer path. Even so, the region would still be faced with severe water scarcity.[24]

There is still a lot of waste in China's use of water, even in water-scarce cities such as Beijing. Because water is supplied free or is greatly underpriced, it is used inefficiently. If Beijing's leaders fail to deal with this issue and are forced to ration water, they will have difficulty creating the conditions that will attract the multinational corporations, luxury hotels, and housing developments that will bring an expanding community from abroad that is desired by the city's leaders.[25]

Finally, there is the as yet incalculable but potentially enormous toll of global warming. Even a modest loss of rainfall or increase in evaporation could disrupt China's finely tuned, highly productive agriculture.

There can be little doubt that China is facing not only a massive conversion of farmland to nonfarm uses if it maintains its breakneck pace of industrialization, but also a heavy diversion of irrigation water to both industrial and residential uses. The bottom line is that this is likely to accelerate the long-term decline in grain production now in prospect for China.

6

Raising Cropland Productivity

In a country where cropland area is no longer expanding, as in China, future growth in food output can come only from raising land productivity. The central question, therefore, is whether China's farmers can raise grain yield per hectare fast enough to offset the inevitable loss of cropland. Data for the four years since 1990 indicate they are losing the race, just as farmers in Japan, South Korean, and Taiwan did two or three decades earlier.[1]

For China, improving productivity essentially means raising yields of the three crops that occupy most of its cropland: rice, wheat, and corn. Each of these three grains accounts for roughly 100 million tons of the 340-million-ton annual grain harvest. Rice and wheat, of course, are the two national staples, with rice dominating in the south and wheat in the north. Some corn is also consumed as food, but most of that harvest is now fed to livestock.[2]

Since mid-century, the changes in China's land productivity fall into four distinct periods. The 28-year span from 1950 through the launching of economic reforms in 1978 was marked by a slow but steady rise in land productivity. The only notable interruption to this was 1955–61, a span that encompassed the Great Leap Forward, when yields were actually declining as millions of the country's farmers were diverted from farming to huge construction projects, ranging from backyard steel furnaces to large earthen dams. Once the country had recovered from the chaos and famine associated with the Great Leap Forward, land productivity resumed a gradual but steady rise. Much of the improvement during this period came from the spread of irrigation.[3]

As the large production teams were broken up under the 1978 reforms and farmers were given long-term leases on the land they worked, productivity soared. Between 1977 and 1984, total grain production went from 199 million to 306 million tons—an increase in the harvest of more than half in just seven years, a feat not equalled by any other major food-producing country. Much of this phenomenal growth came from a dramatic rise in the use of fertilizer. Before the reforms, China was using far less fertilizer than was profitable.[4]

Then from 1984 to 1990, the rise in land productivity slowed, as did the growth in fertilizer use. In the fourth period, since 1990, yields have been rising at an even slower pace.[5]

To understand the long-term potential for increasing rice yields in China, it is instructive to look at the experience of Japan—the first country to initiate a steady rise in yields, which began around 1880. After a century of systematically applying ever more advanced technology and using more sophisticated inputs, Japan's rice yields appeared to hit a ceiling in 1984 at about 4.7 tons per hectare. Since then, yields have fluctuated mostly between 4.5 and 4.7 tons per hectare, except in 1993 when bad weather led to a record drop and in 1994, a recovery year, when they climbed above 4.9 tons. (See Figure 6–1.) Even though Japan's support price for rice is six times the world market level, making it extraordinarily profitable to boost land productivity, farmers there have not been able to sustain the historical rise in yields.[6]

Thirty years ago, China's rice yields were scarcely half those of Japan, but by the early nineties they averaged just over 4 tons per hectare, greatly narrowing the gap. If China's cropland area is underreported for tax reasons,

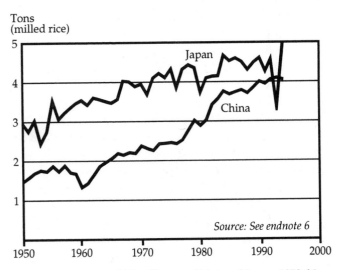

Tons
(milled rice)

Source: See endnote 6

FIGURE 6-1. *Rice Yield Per Hectare, China and Japan, 1950-94*

then yields are overstated, leaving more potential to boost yields.[7]

In the fall of 1994, the International Rice Research Institute in the Philippines, the world's leading center for rice breeding, announced that it had designed a rice variety that would lift yields 20–25 percent above the highest yielding varieties now available in Asia. This gain was achieved by redesigning the plant so that an increased share of its metabolic energy would go into the formation of seed, thus raising the amount of grain produced per hectare.[8]

This new prototype is now being distributed to the major rice-producing countries of Asia. In each country, it will be bred with local varieties to incorporate resistance to indigenous insects and plant diseases and to otherwise adapt it to local growing conditions. If the new variety is successful in China and boosts yields by the maximum theoretical amount of 25 percent, it would add some 30 million tons of rice to the harvest, enough to cover roughly one tenth of the projected growth in demand for grain between now and 2030.

With wheat, perhaps the best way to evaluate China's performance is to compare it with the United States, the other major producer. From 1950 until 1980, wheat yields in the United States were well above those in China. But the agricultural reforms of 1978 lifted China's yields of wheat, much of it grown on irrigated land, above those of the United States. Once they had surpassed the U.S. level in 1982, yields continued to rise, widening their lead. (See Figure 6–2.)[9]

Wheat yields in China climbed 81 percent from 1975 to 1984, jumping from 1.64 tons per hectare to 2.97 tons—a remarkable gain for such a large country. During the following 10 years, however, they rose only 17

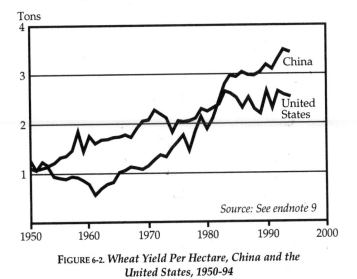

FIGURE 6-2. *Wheat Yield Per Hectare, China and the United States, 1950-94*

percent, reaching 3.48 tons per hectare in 1994. With water scarcity reducing irrigation application rates in the north, where wheat is grown, raising yields will become progressively more difficult.[10]

With corn, the third major grain, China has made impressive progress in recent years. Even so, yields are still scarcely 60 percent those of the United States. This gap could narrow further somewhat, but few countries can approach the yields achieved in the U.S. Corn Belt, which has an ideal combination of deep soils; near-optimal rainfall, both in annual amount and seasonal distribution; temperature; and day length.[11]

As noted earlier, the history since mid-century of grain yield per hectare in China breaks into four distinct periods. (See Table 6–1.) The first, and by far the longest, was from 1950 to 1977. During this period, produc-

TABLE 6-1. *Increase in Grain Yield Per Hectare in China, Selected Periods, 1950–94*

Year	Tons	Increase	Increase Per Year
	(yield)	(percent)	(percent)
1950	1.04		
1977	2.11	+ 103	+ 2.7
1984	3.41	+ 62	+ 7.1
1990	3.77	+ 11	+ 1.8
1994	3.88	+ 3	+ 0.7

SOURCE: See endnote 12.

tivity per hectare doubled. One of the keys to this was the expansion of irrigated area along with the adoption of higher yielding varieties of wheat and rice, developments that paralleled those in other countries in Asia that benefited from the Green Revolution.[12]

By far the most dynamic period in China's recent agricultural history occurred from 1977 to 1984, when the nation led the world in raising land productivity. During this time grain yield per hectare rose by 62 percent, a phenomenal annual rate of nearly 7.1 percent. In just seven years the yield per hectare climbed 1.3 tons, compared with less than 1.1 tons during the preceding 27 years. As a result of the 1978 economic reforms, which broke up the production teams and shifted to family farm units, China in effect compressed into a matter of years much of the Agricultural Revolution that had taken a few decades in many countries. The 1978 reforms triggered a dramatic rise in the use of fertilizer, the input that was primarily responsible for the phenomenal yield increases during this period.[13]

After 1984, yields slowed, increasing less than 2 percent a year through 1990. It was becoming much more difficult to improve productivity. Then from 1990 to 1994, yield per hectare rose even more slowly, edging up only 0.7 percent annually.[14]

China has been remarkably successful in raising land productivity mostly because it irrigates such a large share of its cropland. From 1950 to 1994, the yield per hectare increased 3.7-fold. This suggests that China has done a good job of exploiting the technologies available to raise yields of all grains. If new technologies, such as the rice under development in the Philippines, prove successful, they might provide an opportunity for modest additional gains.[15]

But it is unlikely, unless there is some new breakthrough, that yields will be increased dramatically. At one time, there was high hope that biotechnology would create another generation of high-yielding varieties that would greatly increase yields, much as the earlier generation of high-yielding varieties did that were produced by conventional plant breeding. Unfortunately, this hope has faded as 20 years have passed and biotechnology has not led to a single dramatic gain in yield of any grains. (The new rice type from the Philippines was developed using conventional plant breeding techniques.)[16]

Biotechnology should not be dismissed, however, because it is helping develop increased pest resistance in some crops and drought tolerance in others. In some cases, it is speeding up the process of plant breeding and lowering the cost of achieving such research goals as greater insect and disease resistance. Given the lack of any revolutionary breakthroughs in this field or with conventional plant breeding, the prudent position is to assume that there is not likely to be another generation of grain varieties that will double or triple the yields of existing ones.

China now leads the world in fertilizer use, having surpassed the United States—the long-time leader—in 1986. By 1994, China's fertilizer use had climbed to 29.2 million tons, compared with 18.5 million tons in the United States. (See Figure 6–3.)[17]

The U.S. experience in using fertilizer may be useful in assessing the prospects in China. After growing steadily from 4 million tons in 1950 to more than 20 million tons in the late seventies, U.S. fertilizer use actually declined during the eighties and early nineties. During the peak years from 1979 to 1981, it averaged 21 million tons. From 1992 to 1994, the figure was 18.4 million tons, a decline of more than one tenth. With a half-century tradition of raising fertilizer use, U.S. farmers eventually overshot the mark, using more fertilizer than was profitable.[18]

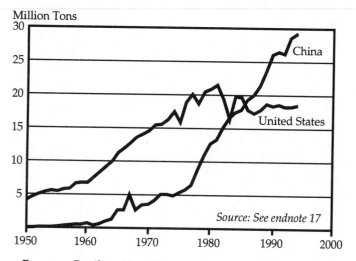

FIGURE 6-3. *Fertilizer Use, China and the United States, 1950-94*

China's use of fertilizer did not reach 1 million tons until the early sixties. It then increased slowly until the agricultural reforms of 1978. Between 1977 and 1986, fertilizer use tripled, accounting for much of the phenomenal growth in grain production during that period. Although it has increased by another 10 million tons since then, the associated growth in the grain harvest has been quite modest, suggesting that China's farmers in the mid-nineties may be using more fertilizer than is economic, much as U.S. farmers were doing at the beginning of the eighties.[19]

One key to assessing the prospects for future trends in fertilizer use is the grain-fertilizer response ratio. In the United States, total grain production divided by fertilizer use yields a response ratio in recent years of roughly 15 tons of grain for each ton of fertilizer used. In China, the ratio is much closer to 11 to 1. This means that fertilizer use in China has reached a point where the response of grain yields is much lower than in the United States, suggesting that fertilizer use is not likely to increase much in the years immediately ahead.[20]

Complicating the assessments of crop yield potential when comparing yields with other countries are the wide variations in inherent land fertility. To the extent that cropland is erodible, it is not surprising that centuries, if not millennia, of cultivation might have reduced the inherent fertility of land. Although comprehensive data on soil erosion are not available for China, as they are in the United States, the data that are available suggest substantial levels of erosion.

Both water and wind erosion take a heavy toll. The Huang He (Yellow River) that drains much of northern China derives its common name from the 1.6 billion tons of topsoil it annually transports to the ocean. And so much of China's topsoil becomes airborne that scientists at the Mauna Loa observatory in Hawaii, the U.S. National Oceanic and Atmospheric Administration's official site for collecting air samples to measure change in atmospheric carbon dioxide levels, can detect the dust within a matter of days after spring plowing starts in northern China.[21]

Soils in China tend to be low in organic matter simply because crop residues—straw, corn stalks, and so on— are typically removed from the field and used either for forage or fuel. In the United States, by contrast, crop residues are usually left on the land to improve soil structure and thus enhance fertility.

One area in which China has an advantage is in nutrient recycling. Many of the nutrients removed from the land in the form of crops are returned in the form of human waste. But even here, the legendary "honeybuckets" and other techniques for recycling sewage onto the land are beginning to break down as labor costs rise and as the society urbanizes.[22]

One trend hanging over China's agricultural future is the growing share of its crop output that is based on the unsustainable use of water. In northern China, for example, where water tables are falling over such a vast area, aquifers will eventually be depleted. (See Chapter 5.) When they are, the irrigation water supply will be reduced as pumping is necessarily reduced to the rate of aquifer recharge. If pumping is double the rate of recharge, for instance, then the eventual depletion of the aquifer will reduce the water pumped by half. Already farmers are losing their water supplies and reverting to rain-fed farming. As this trend continues, it will have the effect not only of lowering yields but also of making harvests more vulnerable to drought.

Air pollution and acid rain are intensifying throughout China as the amount of coal burning increases. One result is lower crop yields and forest productivity not only in China but in Japan and South Korea as well, given the prevailing winds. So far there is no direct measure of how much this reduces yields, but some idea of the extent of damage comes from the United States, where official estimates place the annual crop loss to air pollution at 5 percent or more. It could be as high as 10 percent. If U.S. losses are this heavy, then the more severe air pollution in China is no doubt taking a hefty toll on that country's harvest.[23]

Crop yields in China in the decades ahead will undoubtedly continue to rise, but they are not likely to increase dramatically. Indeed, for some crops, year-to-year changes might be scarcely perceptible. In looking at the two key inputs that have improved land productivity so impressively over the last few decades—irrigation water and fertilizer—there is a growing likelihood that their use may not expand much in the future and it is quite possible that water availability for irrigation will actually decline.

As noted earlier, the central question is whether future rises in land productivity will be sufficient to offset the loss of cropland to nonfarm uses as industrialization progresses. Barring some dramatic new technological breakthroughs, this now seems unlikely.

II

The Shifting World Grain Balance

7

The Growing Grain
Deficit

Estimating China's future food deficit is a scary exercise. Individuals doing the official grain supply projections at the U.N. Food and Agriculture Organization in Rome, at the World Bank in Washington, D.C., and within the Chinese government have been spared some of this trauma simply because they have overlooked the heavy loss of cropland that accompanies industrialization in a country that is already densely populated before the process begins. They have thus assumed that production would continue to climb, closely tracking the rise in consumption, leading to only modest future deficits.[1]

Earlier chapters described what happened to grain production in three other countries that were in this situation—Japan, South Korea, and Taiwan. This chapter simplifies the analysis by drawing primarily on the experience of Japan, the largest of the three.

China today has many similarities with mid-century Japan. Both are densely populated in agronomic terms. Harvested grainland per person in Japan in 1950 was 0.08 hectares, compared with 0.07 hectares in China today. By mid-century, Japan's industrialization was well under way.[2]

To summarize the points made earlier, the main reason for the fall in grain production in Japan during the last four decades has been the loss of grainland to other uses associated with industrialization. In the very early stages, production rose as agriculture intensified and as multiple cropping increased. But before long, industrialization began to claim cropland for construction of factories and warehouses and of the roads and highways that are an integral part of a modern industrial economy. And as nonfarm wages climbed, workers left agriculture, reducing the amount of multiple cropping.

Japan's harvested area of grain peaked in 1955 at 5.1 million hectares. By 1994, it had shrunk to 2.4 million hectares, a decline of just over half. (See Figure 7–1.) This was not the result of a planned effort to reduce the cropland area; it was the consequence of the inescapable need to build thousands of factories, warehouses, and accompanying access roads and of higher wages drawing labor away from farming.[3]

The harvested area also fell as multiple cropping declined. As economic development progressed and wages increased, it became more and more difficult for people farming tiny plots to match the growth in incomes in the industrial sector. The loss of cropland accelerated with industrialization, quickly reaching the point where it overwhelmed rises in land productivity, leading to a

Million Hectares

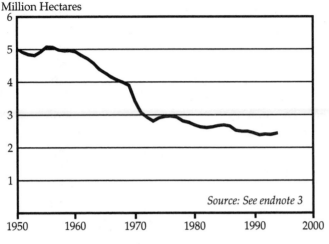

FIGURE 7-1. *Grainland Area in Japan, 1950-94*

steady decline in grain production. Japan's peak production year came in 1960. Despite the hefty rises in rice yield per hectare since then, grain output has fallen 32 percent, nearly 1 percent a year.[4]

From 1950 to 1960, grain production rose more or less in parallel with consumption. (See Figure 7–2.) But then production began its long-term gradual drift downward. Consumption, on the other hand, continued its rapid rise until the mid-eighties, generating an ever wider gap between demand and supply.[5]

This can be seen clearly in the trend in net grain imports. From 4 million tons in 1950, imports increased very slowly until 1960. But once production turned downward, they climbed rapidly until the mid-eighties. This particular graph is for Japan, but the trends are exactly the same for South Korea and Taiwan. The patterns are identical simply because the same forces are at

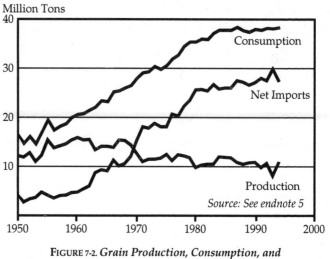

FIGURE 7-2. *Grain Production, Consumption, and Imports in Japan, 1950-94*

work in all three. Only the timing of the production downturn and the associated climb in imports varies, and then only by a matter of years.[6]

From the perspective of the fifties, it was not clear from consumption and production trends that Japan was on the verge of an explosive growth in its dependence on imported grain. But over the next 27 years, grain imports climbed to 28 million tons.[7]

In terms of reliance on imports, Japan depended on grain from other countries for 25 percent of its grain consumption in 1950; by 1985, this had risen to more than 70 percent, where it has since remained. Since 1985, imports have accounted for roughly 72 percent of consumption, except in 1993, when weather sharply reduced the rice harvest and pushed the figure to 76 percent. Most of the growth in the share of grain imported occurred between 1960 and 1975. (See Figure 7–3.)[8]

By the mid-eighties, Japan's grain consumption was levelling off, in part because further gains in income no longer led to large increases in consumption of livestock products. To the extent that consumption of beef, pork, poultry, and other livestock products is still increasing, it is being satisfied in part by imports. For example, although Japan's consumption of beef is relatively low by international standards, the nation now imports more than half of its supply, most of it from Australia and the United States.[9]

China in the mid-nineties may now be where Japan was in the early sixties. Production of grain from 1990 to 1994 has been static, showing little movement up or down. This loss of momentum could mean that a decline in output is imminent. Although production of

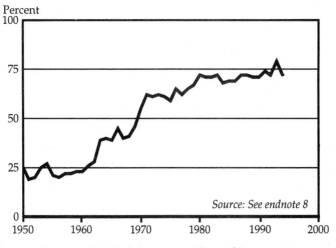

FIGURE 7-3. *Net Grain Imports as Share of Consumption in Japan, 1950-94*

wheat and corn are up slightly, that of rice—which is concentrated in the south, where industrialization is most rapid—has dropped by more than 8 percent between 1990 and 1994. If industrialization continues and spreads more into the central and northern provinces, as government policy is fostering, then production of corn and wheat may soon follow rice downward.[10]

There are several key reasons why China's grain deficit could grow even faster than that of Japan. One is the lack of the seafood option, which enabled Japan to turn to the oceans to satisfy much of its growing demand for animal protein. As noted earlier, if the average Chinese were to consume as much seafood as the average Japanese, China would need the entire world fish catch of 100 million tons.[11]

Because Japan does rely so heavily on seafood for animal protein, the situation there understates the amount of grain that China would need to achieve the same level of animal protein intake. If Japan's 10 million tons of seafood were replaced with pork and poultry, the nation would require roughly another 30 million tons of grain, thus increasing imports from 28 million tons in 1994 to 58 million tons, or 84 percent of a much greater grain consumption.[12]

The second unfavorable situation affecting China that Japan did not have to face is the growing diversion of irrigation water to urban and industrial uses. (See Chapter 5.) Water tables are now falling under much of northern China. The prospect of a reduction in water pumping combined with the diversion of a larger share of this water to urban and industrial uses means that China's grain production could decline over the next few decades even faster than Japan's did.[13]

Yet another unfavorable contrast between China

today and Japan a generation ago is that China has higher grain yields as its rapid industrialization gets under way than Japan did when it entered a similar period of intense industrialization. The unrealized yield potential of each of the three major grains—rice, wheat, and corn—in China today is likely to be much less than it was in Japan at a comparable stage of industrial development.

Even allowing for some boosts in rice yields from research currently under way in China and from the new varietal prototype under development at the International Rice Research Institute, the long-term prospect for large additional rice yield increases in China is not bright. The plateauing in grain production during the last four years suggests that the loss of cropland is already offsetting modest gains in land productivity. As noted in Chapter 4, between 1990 and 1994 China's grain harvested area shrank by just over 1 percent a year. With grain yield per acre rising by just under 1 percent annually, production is declining slightly. Initial estimates put the 1995 grain harvest at 337 million tons, 1 percent below the 341 million tons of 1990. This suggests that the long-term decline in grain output that accompanied industrialization in Japan, South Korea, and Taiwan may be starting in China.[14]

Once the decline in Japan's grain production got under way after 1960, output fell by roughly 1 percent a year, for some 32 percent in total. In looking ahead to 2030, a conservative assumption would be that China's grain production would fall by at least one fifth. If this happens, the 1990 harvest of 340 million tons would fall to 272 million tons by 2030. Again, it should be emphasized that this projected decline may be conservative, given the prospective heavy loss of both cropland and irrigation water to the nonfarm sector.[15]

In projecting demand, two scenarios can be considered. Under the first, demand increases only as a result of population growth and there are no further rises in per capita consumption of meat, milk, eggs, beer, or other food products dependent on the use of grain. The second scenario assumes that the Chinese people continue their recent move up the food chain, albeit at a much slower rate than that of Japan. It suggests that the current annual consumption of just under 300 kilograms of grain per person, including both that consumed directly and that consumed indirectly in the form of livestock products and alcoholic beverages, will increase to 400 kilograms by the year 2030.[16]

Such a consumption level would be roughly the same as Taiwan today, or half the U.S. per capita grain use of more than 800 kilograms per year. This would, in effect, represent a substantial slowdown in the rise in grain consumption per person compared with the 16 years since 1978, when the economic reforms were launched. This is to be expected, since the most rapid rises in consumption come when incomes are rising through the lower ranges rather than the upper ones.[17]

Under both scenarios, the resulting grain deficit is huge, many times that of Japan—currently the world's largest grain importer. In 1990, China produced 340 million tons of grain and consumed 346 million tons, with the difference covered by imports of just 6 million tons. In the first scenario, allowing only for the projected population increase, China's demand for grain would increase to 479 million tons in 2030. In other words, even if China's booming economy produced no gains in consumption per person of meat, eggs, and beer, a projected 20-percent drop in grain production to 272 million tons would leave a shortfall of 207 million tons—roughly equal to the world's entire 1994 grain exports of more than 200 million tons. (See Figure 7–4.)[18]

But China's newly affluent millions will of course not be content to forgo further increases in consumption of livestock products. If per capita grain consumption climbs to 400 kilograms in the year 2030, total demand for grain will reach a staggering 641 million tons. (See Figure 7–5.) Under this scenario, the import deficit would reach 369 million tons, nearly double current world grain exports.[19]

As a reality check to gauge the reasonableness of these figures, a back-of-the-envelope calculation provides a look at what China's grain imports will be in 2030 if its dependence on imports is similar to that of Japan, South Korea, and Taiwan today. Japan, the largest of

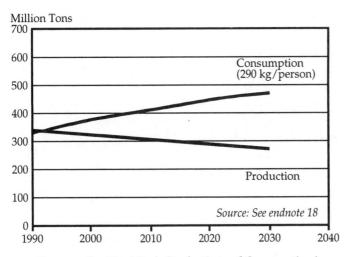

FIGURE 7-4. *Projected Grain Production and Consumption in China Based on Population Growth, 1990-2030*

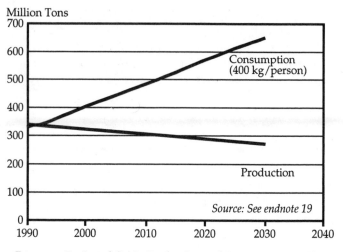

FIGURE 7-5. *Projected Grain Production and Consumption in China Based on Population and Income Growth, 1990-2030*

the three, has one tenth of China's population. Multiplying its 28 million tons of grain imports by 10 produces an import figure for China in 2030 of 280 million tons. Using South Korea as a basis, with a population that is only 3 percent that of China's, yields a figure of 363 million tons of imports in 2030. And Taiwan has a population that is less than 2 percent that of China's, so adjusting its 6 million tons of imports per year translates into 333 million tons of grain.[20]

While China's political leaders are reluctant to recognize the possibility of a grain deficit so large that it could lead to global food scarcity, at least one Chinese scientist has made calculations similar to these. Professor Zhou Guangzhao, head of the Chinese Academy of Sciences, observes that if consumption per person nationwide reaches the level of the most affluent coastal provinces (about the same level as Taiwan) and if the nation continues to squander its farmland and water resources in a breakneck effort to industrialize, "then China will have to import 400 million tons of grain from the world market. And I am afraid, in that case, that all of the grain output of the United States could not meet China's needs."[21]

At this point, there is little doubt that massive grain deficits lie ahead for China. How quickly they will begin to unfold depends more on the rate of industrialization than any other single factor. Continuing rapid industrialization will chew up cropland and reduce grain production while at the same time moving the country's 1.2 billion people—and counting—up the food chain. With consumption climbing and the possibility that production could soon start to fall, a massive grain deficit within a matter of years appears likely.[22]

Such a scenario does not depend on a continuation of the 10–13 percent annual economic growth of the last four years. Existing projections have the economy of China growing at an average of 8 percent a year from 1995 to 2005. The most reasonable assumption would be that this rate would gradually slow as 2030 draws closer until by that time it might well be in the 2–3 percent range, comparable to the average of industrial countries in recent years.[23]

China has always been reluctant to import large quantities of grain, but evidence that official stocks were largely depleted in 1994 in efforts to prevent runaway grain prices suggests that future shortfalls can be filled only by importing. Even though millions of tons of grain were released to the market from the official government grain reserve in 1994, grain prices still climbed throughout the year. In its closing months, grain prices were 60 percent higher than those of a year earlier. With food purchases accounting for close to half of total consumer expenditures, food price rises pushed the overall rate of inflation to 24 percent for 1994, the highest since the Communist party took power in 1949.[24]

Although the government allowed grain prices to increase substantially in order to raise rural incomes and to encourage people to stay on the land, it cannot permit this to continue much longer without jeopardizing the political stability of the cities. From now on, threatened price rises will have to be checked either with rationing, something the government is reluctant to do, or with imports.

The year 1994 appears to have been a pivotal one for China, perhaps marking the transition from an era of being essentially self-sufficient to one of growing dependence on imported grain. In early 1994, China was still exporting substantial amounts of corn, a short-lived surplus from a surge in corn production that occurred in the early nineties. By late spring, private firms were reneging on corn export contracts to South Korea and other countries. In effect, they were arguing that domestic corn prices had risen to the point that they could not procure corn at the contract price. In late summer, corn exports were officially banned.[25]

After having oscillated between small net exports and imports of grain during the last 20 years, China may be about to begin importing steadily expanding amounts. In the trade year 1993/94, it actually had net grain exports of some 8 million tons. In 1994/95, its net grain imports are estimated at 16 million tons. But even the shift of 24 million tons in one year in the grain balance does not reflect the full change, since Beijing released large amounts of grain from stocks as it tried to check the rise in prices. In addition, the price rise itself checked the growth in demand, reducing the actual shortfall to a level much lower than it would have been otherwise.[26]

Although data on government-held grain stocks are not made public, reserves apparently were reduced substantially by releases during 1994. One source reports that between October 1993 and the end of 1994, stocks dropped from 40 million to 26 million tons. Unless grain stocks can be rebuilt, future efforts to check rising prices will likely have to rely heavily on imports.[27]

8

Competition for Grain

Concern about food security runs deep in China. The current leaders, remembering all too clearly the Great Famine, are committed to self-sufficiency in food, at least in their public statements. They are also committed to industrialization—getting rich is now glorious.

It is hard to imagine a government any more committed to industrialization, yet Beijing faces a dilemma. It cannot continue to industrialize and remain self-sufficient in food. In addition, China has decided to build an auto-centered transport system, which will claim a vast amount of additional cropland. If policymakers follow this path and if the frenetic pace of industrialization continues, imports of grain may well reach an unprecedented level before the end of this decade.[1]

In confronting a deficit on the scale projected, two key

questions arise: Will China have enough foreign exchange to import the grain it needs? And will the grain be available? On the first count, if the premise underlying this demand is a continuation of the economic boom, there would likely be ample income from industrial exports to pay for the needed grain imports at current prices.

China's nonagricultural exports are growing by leaps and bounds. Since the economic reforms were launched in 1978, they have grown at a prodigious rate. In 1994, these exports surpassed $100 billion for the first time. Recent record levels of foreign investment by global corporations are designed both to ensure access to the Chinese market and to capitalize on the vast pool of cheap labor within China to ensure competitiveness in world markets. This latter goal translates into growing exports.[2]

If exports continue to grow as expected, China will have ample foreign exchange with which to import grain. Its trade surplus with the United States in 1994 reached nearly $30 billion. (See Table 8–1.) Filling a 100-million-ton import deficit, which is equal to nearly half of world grain exports, by bringing in wheat or corn at 1994 prices of roughly $150 a ton would require $15 billion. Given its trade surplus with the United States alone, China could buy all U.S. grain exports—grain that now goes to more than 120 grain-deficit countries—even if grain prices doubled.[3]

Given the likely continuing growth in China's nonagricultural exports, importing 200 million or even 300 million tons of grain at current prices would be within economic range if the country's leaders were willing to use a share of export earnings for this purpose. Of course, this could mean cutting back on capital goods

TABLE 8-1. *China's Trade Surplus with the United States, 1980–94*

Year	Exports to the United States	Imports from the United States	Difference
	(billion dollars)		
1980	1.1	3.8	− 2.7
1981	1.9	3.6	− 1.7
1982	2.2	2.9	− 0.7
1983	2.2	2.2	0.0
1984	3.1	3.0	+ 0.1
1985	3.9	3.9	0.0
1986	4.8	3.1	+ 1.7
1987	6.3	3.5	+ 2.8
1988	8.5	5.0	+ 3.5
1989	12.0	5.8	+ 6.2
1990	15.2	4.8	+ 10.4
1991	19.0	6.3	+ 12.7
1992	25.7	7.4	+ 18.3
1993	31.5	8.8	+ 22.7
1994	38.9	9.4	+ 29.5

SOURCE: See endnote 3.

imports and possibly on oil imports, which in turn could diminish the inflow of technology and energy needed to sustain rapid economic growth.

The more difficult question posed earlier is, Who could supply grain on this scale? The answer: no one. No one exporting country nor even all of them together can likely expand exports enough to cover more than a small part of this huge additional claim on the world's exportable grain surplus. In the real world, the price of grain would rise, reducing consumption and imports while stimulating production and exports until a new balance was reached.[4]

Looking back at the changing trade balance since

mid-century gives some sense of what the grain export potential might be and how it relates to China's projected needs. Trade is a sensitive indicator, measuring the balance between the supply and demand of a product within a particular country. Since grains are staple foods, it effectively gauges the capacity of a country to feed itself. It is a measure of the difference between the demand for food and the carrying capacity of a country's land and water resources with any given level of technology and investment.

In 1950, North America was the only one of the world's seven regions that had a substantial export surplus of grain. (See Table 8–2.) Indeed, exports from North America of 23 million tons were offset by the 22 million tons of grain imported by Western Europe. Asia was beginning to import, but on a relatively small scale—6 million tons—compared with Europe.[5]

Over time, this pattern changed. By 1990, the world grain trade was still dominated by exports from North

TABLE 8-2. *The Changing Pattern of World Grain Trade, 1950–90*[1]

Region	1950	1960	1970	1980	1990
	(million tons)				
North America	+ 23	+ 39	+ 56	+ 130	+ 110
Western Europe	− 22	− 25	− 22	− 9	+ 27
E. Europe. & Soviet Union	0	0	− 2	− 44	− 35
Latin America	+ 1	0	+ 4	− 15	− 10
Africa	0	− 2	− 4	− 17	− 25
Asia	− 6	− 17	− 37	− 63	− 81
Australia & New Zealand	+ 3	+ 6	+ 8	+ 19	+ 14

[1]Plus sign indicates net exports; minus sign, net imports.

SOURCE: See endnote 5.

America (110 million tons), but imports were dominated by Asia (81 million tons). Latin America had gone from a slight surplus to a net import position of 10 million tons, as grain-buying by Mexico, Brazil, Venezuela, and other smaller countries more than offset exports from Argentina. Eastern Europe and the former Soviet Union emerged as a substantial importer during this period, as did Africa. By 1990, they were bringing in 35 and 25 million tons of grain, respectively.[6]

The big surprise was Western Europe. From the beginning of the Industrial Revolution through 1970, the region was a net importer of grain. But as its population stopped growing, as yield-raising technologies continued to advance, and as subsidies stimulated output, grain production moved above consumption during the eighties, making it a net exporter of 27 million tons in 1990.[7]

These trends in the regional grain trade pattern over the last four decades give some sense of the major shifts occurring in world trade and some of the reasons for it. But to get a clearer sense of whether grain will be available to fill China's looming deficit, it is useful to look at some of the key exporters on a country-by-country basis.

Australia, for example, typically exports 12–15 million tons of grain a year. It does not often go much above that, but it sometimes falls well below it in years of severe drought, such as 1994–95, when it exported only 6 million tons. Australia is unique among industrial countries in its limited progress in raising grain yield per hectare and, therefore, overall grain production. During the last 40 years, its average grain yield has gone up by roughly half, less even than in Africa. Australia is unable to double and triple yields, as so many countries have done, because it is a semiarid or arid country. It has little water, and without water the potential for fertilizer use is limited, thus preventing any striking rises in grain production.[8]

Canada is in a somewhat similar situation. Most of its grain is produced in the northern reaches of the Great Plains, where farmers must contend not only with low rainfall, similar to that in the U.S. Great Plains, but also with a harsh winter and short growing season. While farmers in the U.S. Great Plains largely grow winter wheat, those in Canada can grow only the lower-yielding spring wheat. Without the abundant rainfall of the U.S. Corn Belt, the potential for using fertilizer to boost yields on Canada's grainland is limited.

In addition, Canada is experiencing difficulty even in maintaining its wheat production as the demand for vegetable oils climbs. Several years ago, Canadian plant breeders succeeded in producing a high-quality cooking oil from rapeseed, a traditional oilseed crop that is widely grown in Europe and China. Seeking a more marketable name, the Canadians decided to call rapeseed canola—short for Canadian oil.[9]

To satisfy the soaring demand for canola, which is one of the least saturated of the vegetable oils and therefore in demand for dietary reasons, Canadian farmers had to sacrifice wheatland. As a result, Canada's wheat crop fell by a fifth from 1990 to 1994 as canola production nearly tripled. This dramatic rise in output pushed the value of the canola harvest up to that of wheat in 1994, marking perhaps the first time in Canadian economic history that wheat was not the dominant crop. Given the desire for more vegetable oil, particularly among low-income consumers in China and India whose incomes are rising rapidly, it seems likely that the global demand for this product will continue its recent rapid climb.[10]

Another traditional grain-exporting country, Argentina, faces a similar conflict between grain and oilseeds. Here it is soybeans that compete with grain in the use of land. Among the exporters, however, Argentina almost certainly has the largest share of unused or underused production potential. If it can realign its economic policies to remove the heavy export tax on farm commodities, Argentina has the agronomic potential to perhaps double its annual grain exports from the recent 10–12 million tons. Large though such a gain might be for Argentina, it appears small in light of China's projected needs.[11]

If prices rise high enough to provide sufficient reimbursement to Europe's farmers, the European Union might return to production the 12.5 percent of the grainland that it set aside in 1995 under the common agricultural policy. If this happens, Europe could continue to export grain at a level comparable to that of recent years. Otherwise, with the dismantling of the high support prices, Western Europe's grain production may well fall, reducing its exportable surplus.[12]

Eastern Europe and the former Soviet Union are still importing grain in substantial amounts. But if the new republics could reverse the recent decline in grain production associated with economic reforms, as the East European countries did beginning two years ago, they could again export grain. Russia, Ukraine, and Belarus, all with stable populations, could export at least modest quantities of grain. Partially offsetting this potential surplus are the Asian republics, such as Kyrgyzstan, Uzbekistan, Turkmenistan, and Tajikistan, which have population growth rates well above that of India. This, combined with water scarcity, makes growing food deficits in these countries appear inevitable.[13]

Thailand has also traditionally exported rice and corn, but it has recently lost its exportable surplus of corn as population and rising incomes have driven the demand for livestock products upward. It should still be able to maintain exports of rice for some time, however, perhaps at roughly the current level of 5 million tons per year.[14]

The only other major grain exporter is the United States, which can produce and export more grain than it now does—but probably not nearly as much as some people think. In 1994, the United States returned to production all the grainland idled under commodity supply management programs. Even with this land in use and one of the best U.S. harvests in memory, world grain stocks still fell.[15]

The one other reserve that the country can call on is the 14 million hectares (36 million acres) in the Conservation Reserve Program (CRP). Located mostly in the wheat-growing region of the western plains, this land typically produces 2.5 tons of grain per hectare (roughly 35 bushels per acre), compared with Midwestern cornland that produces 8 tons per hectare (120 bushels per acre).[16]

Some of the CRP land, all of which is highly erodible, could be farmed on a sustainable basis if agricultural practices were adjusted using, for example, minimal tillage practices, crop rotations, or some other soil-conserving practices. Part of the CRP land is so erodible that it should be left in grass and grazed, which is the only use that is sustainable.

Two key questions in the United States are, How much cropland will be available? And how much irrigation water will be available over the longer term? According to the U.S. Department of Agriculture, some 21 percent of the irrigated cropland in the United States is watered by drawing down underground water tables. Much of this irrigation is in the central and southern Great Plains, where farmers irrigate with wells and overhead sprinklers, drawing water from the Ogallala aquifer, which is essentially a fossil aquifer. In the more shallow southern end, the aquifer has already been depleted in some locations and farmers are going back to rain-fed farming. This is most common in the Texas panhandle region and is largely responsible for the decline in irrigated area in Texas of some 30 percent in the last 15 years.[17]

In looking ahead at the U.S. potential during the next four decades, a combination of Ogallala depletion and the diversion of irrigation water to satisfy the demands of Sun Belt cities, such as Los Angeles, Phoenix, Tucson, Las Vegas, El Paso, and Denver, will further reduce the water available for irrigation in the southern Great Plains and the Southwest. The growth in U.S. irrigated cropland shows a clear slowdown in recent years as the demand for water has collided with the sustainable yield of aquifers and as depletion of the Ogallala aquifer begins to shrink irrigation water availability. (See Figure 8-1.)[18]

A second constraint on the growth in U.S. food production is the loss of cropland to nonfarm uses. Between 1990 and 2030, the United States is projected to add 95 million people. Satisfying the nonfarm land needs of this 38-percent addition to U.S. population means sacrificing at least some cropland. With the U.S. average of three people per household, this means building 32 million housing units—freestanding houses, townhouses, or apartments. As in the past, a large share of these will

Million Hectares

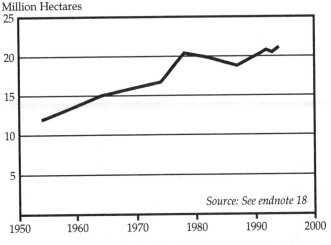

FIGURE 8-1. *Irrigated Land in the United States, 1954-94*

be built on cropland. Beyond housing, these new Americans will also require a proportionate increase in factories, schools, churches, shopping malls, golf courses, cars, roads, and parking lots.[19]

With the additional 95 million people comes a simultaneous jump in the demand for grain of some 76 million tons—assuming today's consumption of 800 kilograms per person annually—along with loss of cropland. The net effect will make it more difficult to expand the exportable surplus of grain.[20]

In looking at the longer term U.S. grain export potential, there is lurking in the background the possibility of another summer, or perhaps even a series of summers, like that of 1988—when severe heat and drought dropped the U.S. grain harvest below consumption for the first time in history. Fortunately, the United States was then holding huge, near-record carryover stocks of grain. By exporting this reserve, it met the import needs of the more than 120 grain-importing countries that look to the United States for part of their grain supply. In 1995, world carryover stocks of grain are at the lowest level in 20 years.[21]

If the United States were to go through a summer like that of 1988 at a time when grain stocks are so low, world grain markets would soon be in chaos. If the buildup in greenhouse gases continues unabated, climate models show that temperature rises in the interior of the continent, specifically the corn-growing Midwest and the Great Plains, will be disproportionately great. If these models are close to the mark, it may be only a matter of time until rising temperatures begin to affect the capacity of the United States to supply grain.[22]

Another important question, of course, is how much the United States can raise land productivity above current levels. With wheat, the rise in yields has slowed markedly during the last decade. (See Chapter 6.) Corn yields have continued to rise, but are reaching a level where it may be difficult to sustain steady increases. The bottom line is that U.S. farmers could produce somewhat more than they now do on a sustainable basis if a market for their output existed.

World grain exports grew rapidly during most of the last 34 years, levelling off only during the last decade. (See Figure 8–2.) To some extent, this is the result of limited growth in import demands.[23]

Thus far this analysis has focused on grain, but China is also importing growing amounts of vegetable oil, sugar, and cotton. As noted in Chapter 3, China is expected to import 3.5 million of the 9.3 million tons of vegetable oil it consumes in 1995. Dependence on imports as a share of consumption has climbed from

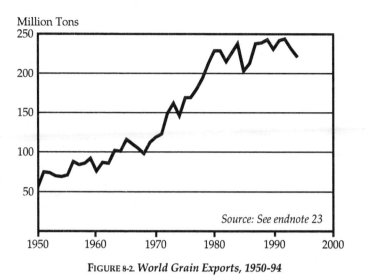

Million Tons

FIGURE 8-2. *World Grain Exports, 1950-94*

Source: See endnote 23

zero in 1984 to 38 percent in 1995.[24]

Sugar consumption in China, which has one of the world's lowest levels of per capita consumption, is going up by 6 percent per year. Imports are also climbing. As recently as 1991, China was a net exporter of sugar. Within three years, the nation went from being more than self-sufficient to depending on the outside world for 17 percent of its sugar.[25]

A similar situation exists for cotton. As recently as 1987, China was a major net exporter of cotton. In 1995, it is expected to import 3.2 million bales net against total use of 20 million bales. This 16 percent of consumption is expected to rise steadily in the years ahead as the growth in China's textile industry continues to outstrip that of the cotton harvest.[26]

Anyone following commodity markets has to be impressed with the extent to which decisions by China to import are affecting these markets. Typical of these is a statement appearing in the *Wall Street Journal* in late 1994: "World sugar prices surged to their highest levels in more than four years as rumor that China is poised to make a major purchase raised the prospect of even greater demand and tighter supplies."[27]

China's heavy cotton imports in early 1995 have helped push world cotton prices to the highest level since the U.S. Civil War. Despite a record world soybean harvest and the largest year-to-year increase ever in the world crop, vegetable oil prices actually rose following that harvest, in large part because of China's record imports. With such high world cotton and vegetable oil prices, farmers everywhere will be tempted to shift resources from grain into these crops, making it even more difficult to expand the world grain harvest. U.S. farmers expanded the area planted to cotton in 1995 by 18 percent, making it the largest planting in a generation.[28]

Even as China is facing the potential need for massive imports of grain, many other countries are in a somewhat similar situation. Ten of the more populous developing countries that are projected to add substantial numbers of people over the next four decades are included in Table 8–3. Allowing only for increases in consumption associated with population growth, which means no improvement in diet, many of them will experience a doubling or tripling of demand. Among those projected to triple their populations are Iran, Nigeria, and Ethiopia. Pakistan's population is projected to nearly triple. Countries that are projected to double include Bangladesh and Egypt.[29]

On the supply side, almost all these countries face land and water constraints, some of them severe. Nonetheless, it is assumed that production in each of these

TABLE 8-3. *Grain Imports for Selected Countries, 1990,*
With Projections for 2030

Country	1990	2030
	(million tons)	
India	0	− 45
Bangladesh	− 1	− 9
Indonesia	− 3	− 12
Iran	− 6	− 32
Pakistan	− 1	− 26
Egypt	− 8	− 21
Ethiopia & Eritrea	− 1	− 9
Nigeria	0	− 15
Brazil	− 6	− 4
Mexico	− 6	− 19
	− 32	− 190

SOURCE: See endnote 29.

countries will increase by roughly half or more and, in some countries, that it will more than double. Yet huge deficits are in prospect. In 1990, this group of 10 countries imported 32 million tons of grain, roughly one sixth of the world total. For comparison purposes, collectively they imported only slightly more than Japan did. By 2030, these countries—assuming no improvement in diet—will need to import 190 million tons of grain. This is six times the amount they import today and nearly equal to total world grain exports in 1994.[30]

The point of these projections is that competition for grain imports in the years ahead is likely to intensify dramatically even without China's emergence as a massive importer. This suggests that the world grain market soon will be converted from a buyer's to a seller's market. From mid-century onward, exporting countries always seemed to be competing for markets that were never quite large enough. This buyer's market has been dominant except for a brief period in the mid-seventies, when grain prices soared following a secretive Soviet wheat purchase when Moscow decided to offset a harvest shortfall with imports rather than belt-tightening.[31]

Not only have grain exporters faced a buyer's market, but the strong competition among exporting countries to produce ever more efficiently lowered the real price of grain substantially from mid-century through the early nineties. Except for the brief period in the mid-seventies, the real price of wheat and rice has been declining throughout this period. (See Figure 8-3.)[32]

This has created an ideal environment for alleviating hunger. Even low-income countries with limited foreign exchange were faced with a gradually declining outlay for grain imports. Fixed budgets for food aid provided

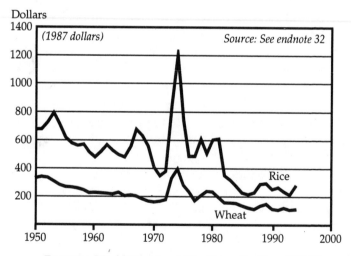

FIGURE 8-3. *World Wheat and Rice Prices Per Ton, 1950-94*

for steadily growing quantities as the real price declined. Unfortunately, however, little progress has been made in ridding the world of hunger.

If this buyer's market is converted into a seller's market, importing countries will soon find themselves competing vigorously for supplies of grain that never seem adequate. In such a world, the politics of scarcity will replace the politics of surplus, bringing the risk of grain export embargoes in countries trying to control inflationary food price rises.

III

Facing Scarcity

9

Entering a New Era

As the world progresses through the nineties, each year brings additional evidence that we are entering a new era, one quite different from the last four decades. An age of relative food abundance is being replaced by one of scarcity. As the one fifth of humanity who live in China seek to join the affluent one fifth already living high on the food chain, the transition into the new era will be accelerated.

There is no historical precedent by which to assess the effects on the world food economy of China's projected emergence as a heavy importer of grain. Although the 13 million people currently added to China's population each year are expanding the demand for food, most of the growth comes from the extraordinary rise in incomes. Since the economic reforms were launched in

1978, China's economy has expanded fourfold. In economic terms, there are now four Chinas where there was only one 17 years ago.[1]

If China continues on its current industrial track, it will reach a level of industrialization by 2030 similar to that today in its more industrialized neighbors—Japan, South Korea, and Taiwan. Its dependence on grain imports would also be similar, for reasons outlined in earlier chapters.

China's leaders publicly argue that they will not become dependent on imported grain, that China will always feed itself. But this is highly unlikely. They argue that they will protect their cropland from conversion to nonfarm uses, but it is difficult to imagine any country making a greater effort to do this than Japan, yet that nation has lost half of its grainland since 1950. Some argue that China will increase the productivity of its land so much that it will not have to import grain. But China's yields are already quite high by international standards—with rice yields close to those of Japan and wheat yields well above those of the United States. The bottom line is that if China continues its rapid industrialization, it will inevitably be forced to import heavily to satisfy future food demand.[2]

Why do China's leaders stoutly maintain that they will never become dependent on imported grain? Perhaps they genuinely believe that the phenomenal rises in land productivity that followed the economic reforms of 1978 will somehow be sustained indefinitely into the future. Or it could be that they are simply practicing a form of denial. For the Chinese who barely survived the Great Famine, it must be unbelievably difficult to accept the insecurity that is associated with becoming heavily dependent on the outside world for food.

Although the projections in Chapter 7 show China

importing vast amounts, movements of grain on this scale are never likely to materialize simply because they, along with climbing import needs for other countries, will overwhelm the export capacity of the small handful of countries with an exportable surplus. As China's grain imports rise, they are likely to drive prices upward, making it increasingly costly to import on the scale projected. The price rise associated with the growing imbalance could lead to economic disruption everywhere and, in low-income countries, to political instability that would dwarf the effects of the oil price hikes in the seventies. There are, after all, substitutes for oil. But there are no substitutes for food.

China can reduce its growing dependence on imported foods in a number of ways. One is to stabilize population before it reaches the peak of 1.66 billion projected for 2045. If, for example, the one-child policy were now extended to minority groups, which make up a substantial share of China's 1.2 billion people, and if incentives to adhere to this goal overall were strengthened, this could help lower the level at which population will peak before it begins to decline.[3]

Beijing can also alter its industrial policies in a way that will save cropland. At present, it is simply adopting the western industrial model, emphasizing, for instance, the development of an automobile-centered transportation system that will inevitably consume vast amounts of cropland. Alternatively, China could concentrate on developing a modern national rail transport system for both freight and passengers that would be state-of-the-art and would need only a fraction of the land. For passengers, the bike-to-train commute system that is most highly developed in Japan and the Netherlands provides an appropriate model.

In coping with water scarcity, there are innumerable

opportunities for increasing efficiency. The most important single step here is to begin charging the full cost of supplying water, abandoning the subsidies that lead to wasteful use. Even in Beijing, where water scarcity threatens future development, urban dwellers get water at a fraction of its real cost.[4]

One senior Chinese official has publicly raised doubts as to whether it is in the country's interest to keep moving up the food chain. Such questions notwithstanding, societies move down the food chain only when they are forced to do so by food rationing or by rising food prices.[5]

It is an accident of history that China is turning to world markets just when the growing world demand for food is colliding with the sustainable yield of oceanic fisheries, the sustainable yield of aquifers in key food-producing regions, and the limits of available crop varieties to respond to additional applications of fertilizer. The net effect of these constraints is slower growth in world food production.

In the past, rising prices have typically spurred greater growth in output. If seafood prices rose, then fishers would simply invest in more trawlers and expand the catch. That approach today would only hasten the collapse of oceanic fisheries. Before, when grain prices went up, farmers would invest more in irrigation wells. Unfortunately, in much of the world today that response will just accelerate the depletion of aquifers. In the past, rising grain prices stimulated farmers to apply larger amounts of fertilizer. Today, using more fertilizer in many countries has little effect on production.

Neither fishers nor farmers have been able to keep up with the growth in population in recent years. The shift from an era in which both were expanding output much faster than population to one in which both are trailing population growth was well under way before China's prospective emergence as a large importer. But as China turns to the outside world for more grain, it will accelerate the transition from an era of surpluses to one of scarcity.

After four decades of spectacular progress in expanding world food production, many people thought that the steady growth in both oceanic and land-based food production would continue more or less indefinitely. Among those making this assumption were those responsible for making world food supply and demand projections at the U.N. Food and Agriculture Organization and at the World Bank. One result of the simple extrapolation exercise they engage in is that they yield "no problem" projections, thus lulling the world into a false sense of complacency.[6]

A convenient way of comparing the old and the new era is simply to look at the data from 1950 to 1990, which roughly brackets the old era, with the projections for 1990 to 2030, which is the era policymakers must now focus on. The fish catch, for example, grew from 22 million tons in 1950 to 100 million tons in 1990, a 4.6-fold increase. During the next 40 years, it is not expected to increase at all. (See Table 9–1.)[7]

With grain output, production went from 631 million tons in 1950 to 1,780 million tons in 1990, an increase of 1,149 million tons. During most of this period, the growth in grain production outstripped that of population, boosting per capita grain output by some 40 percent between 1950 and 1984. Since then, growth in grain production has slowed.[8]

In contrast to the last four decades, when farmers boosted output by some 28 million tons a year, our pro-

TABLE 9-1. *World Seafood Catch and Grain Output, 1950–90, With Projections to 2030*

Commodity	1950	1990	2030	Change 1950–90	Projected Change 1990–2030
			(million tons)		
Seafood Catch	22	100	100	+78	0
Grain Output	631	1,780	2,149	+1,149	+369

SOURCE: See endnote 7.

jections show grain production growing only 369 million tons in the next four decades, or 9 million tons a year. This projection takes into account the diminishing response of crop yields to additional fertilizer use; the growing scarcity of fresh water; the shrinking backlog of unused agricultural technology; the social disintegration occurring in some countries, mostly in Africa; and the heavy loss of cropland occurring in Asia.

The projection assumes slower growth in production almost everywhere, and in some countries—notably China—actual declines. If grain production is declining in China—a country that now accounts for one fifth of the world grain harvest—by roughly 1 percent a year, as projected, it will take a substantial gain elsewhere just to offset this loss.

Four harvests into the new era, there is ample reason to take these projections seriously. Since 1990, there has been no growth in the world grain harvest. Indeed, the 1994 grain harvest is actually smaller than that of 1990. Similarly, since 1990, the world fish catch has not increased at all. While four years does not determine a trend for the next four decades, this lack of expansion is a matter of concern. The projected growth in grain production from 1990 to 2030 of 9 million tons per year is less than a third of the 28-million-ton annual growth between 1950 and 1990. Nonetheless, during the last four years, even this has not been realized.[9]

Between 1987 and 1994, world carryover stocks of grain were reduced from an all-time high of 465 milion tons to 298 million tons, dropping stocks to the lowest level in 20 years. Without the addition to world grain supplies of these nearly 24 million tons of grain per year, grain prices would likely have risen sharply.[10]

The new era we are moving into will be so different that many of our traditional reference points will be lost. (See Table 9–2.) The rise in grain output per person and seafood catch per person that has been under way since mid-century will be replaced by declines of both. Instead of dropping in real terms, food prices will be rising, as we already see with seafood. Rice scarcity is likely to be next simply because its production is constrained by the availability not only of land but also of fresh water. China's rice harvest in 1994 was 8 percent less than in 1990 not because of failed technology, but because of the heavy loss of riceland in its southern provinces. Price rises for rice will probably be followed by those for wheat and then other grains.[11]

The era of scarcity that lies ahead is currently most evident with seafood, where human demand is pressing against the sustainable yield of oceanic fisheries. In recent years, seafood prices have been rising by 4 percent a year in real terms.[12]

Thirty years ago in the United States, poor people who could not afford meat ate fish. Today seafood costs more than most types of meat. What is happening to

TABLE 9-2. *Food and Economic Indicators in Two Eras*

Indicator	Economic Era 1950–90	Environmental Era 1990–2030
Seafood catch per person	rising	falling
Grain production per person	rising	falling
Food prices	falling	rising
Grain market	buyer's market	seller's market
Politics of food	dominated by surpluses	dominated by scarcity
Income per person	rising	may decline for much of the world

SOURCE: See endnote 11.

seafood prices is instructive as it provides a glimpse of how future scarcity can affect prices of other foodstuffs. With seafood prices, we are seeing the economic manifestations of our failure to stabilize population before reaching the limits of the ocean's carrying capacity. Unfortunately, we may see price rises for other foodstuffs as similar imbalances between supply and demand begin to develop.

In effect, the era from 1950 to 1990 was shaped by the pursuit of economic growth, including—significantly—record growth in world food production. During most of this period, there were no serious natural constraints on the growth in the seafood catch or in grain production, only economic ones.

The environmental era, however, will be characterized by a collision between economic trends and the earth's many natural limits, such as the sustainable catch of oceanic fisheries or the sustainable yield of aquifers. Increasingly, economic trends and, indeed, the evolution of the global economy will be shaped by environmental issues, such as the need to use water more efficiently or the need to stabilize climate.

One consequence of the increasingly frequent collisions between growing human demands and limits on the earth's natural systems is likely to be a further slowdown in world economic growth. (See Table 9–3.) Although the food-producing sector is a small share of the global economy, it is so basic that any difficulties in adequately expanding output are likely to cause economic disruption and political instability.[13]

A review of global economic growth since mid-century shows growth peaking during the sixties at an annual rate of 5.2 percent, then dropping in each of the next two decades. During the first four years of this decade, it has averaged only 1.4 percent, slightly less than population growth. This means that thus far during this

TABLE 9-3. *World Economic Growth by Decade, Total and Per Person*

Decade	Annual Growth	Annual Growth Per Person
	(percent)	
1950–60	4.9	3.1
1960–70	5.2	3.2
1970–80	3.4	1.6
1980–90	2.9	1.1
1990–94 (prel.)	1.4	− 0.3

SOURCE: See endnote 13.

decade, income per person has actually declined slightly.[14]

At issue is how the world economy will be affected if the growth in food production continues to lose momentum. With the continuing loss of cropland not only in China but throughout Asia, a densely populated region that contains at least half the world's people, it will become progressively more difficult to achieve rapid growth in world food output.

The levelling off of the world fish catch and the loss of momentum in grain production overall could further undermine economic expansion. If it does, the world could end this decade with an actual decline in income per person, following the path forged by Africa during the eighties.[15]

10

Priorities in an Era of Scarcity

When Rachel Carson wrote *Silent Spring*, she launched the modern environmental movement, a transformation that would envelop the entire world. Her warning of the threat that residual pesticides, such as DDT, posed to bird populations led to quick actions and gave environmental concerns a pollution focus, one that has dominated the movement ever since. Pollution of air and water continue to be important issues, particularly because of the effect of pollutants on human health. But an even more fundamental concern—the earth's capacity to produce enough food to satisfy our expanding demand—is now emerging as the overriding environmental issue as the world approaches the twenty-first century.

The loss of food security promises to become the defining focus of the global environmental threat. This is most evident with the oceanic food system, where human demand is pressing against the limits of natural fisheries. These limits mean that the seafood catch per person will shrink each year as long as population growth continues.

As noted in earlier chapters, spreading water scarcity, the limits of available crop varieties to effectively use more fertilizer, and the lack of fertile new land to cultivate are combining to slow the growth in food production on land. After expanding much faster than population from 1950 to 1984, grain production has since fallen behind population growth, dropping nearly 1 percent per year. There are still opportunities to expand grain production, but none promise the quantum jumps in output that came with earlier advances, such as the hybridization of corn or the growth in irrigation, from mid-century forward.[1]

Rising prices, already dramatically evident with seafood, are likely to spread to rice, wheat, and other food staples, making survival an issue for the world's poor. At the national level, food scarcity will affect economic stability and, for some governments, perhaps political survival.

For the first time, an environmental event—the collision of expanding human demand with some of the earth's natural limits—will have an economic impact that affects the entire world. Rising food prices will touch all of us one way or another.

As the world contemplates the prospect of scarcity, it must also face the issue of distribution. As long as the pie was expanding more rapidly than population was growing, political leaders could always urge the poor to be patient because eventually their share would also rise.

If the food supply is not expanding at all, as with sea-food, or much more slowly than population, as with grain, the question of how the pie is divided becomes a much more immediate political issue.

One way of distributing scarce resources is to let the market do its job. Indeed, given the economic reforms in the former Soviet Union and China, reliance on the market to distribute food is now nearly worldwide. Whenever demand outruns supply, the price rises, reducing demand while encouraging additional supply. From a purely economic standpoint, the market does a good job of balancing demand and supply and distributing food. But from a social point of view, rising prices of food can quickly become life-threatening for the world's poorest. For the Third World's rural landless and its shantytown residents who already may spend 70 percent of their income on food, even a modest rise in food prices can threaten survival. China's prospective emergence as a massive importer of food may well force the world to address this long-ignored issue of distribution.

If grain prices rise in the years ahead, as now seems likely, they could create an unprecedented degree of insecurity. No economic indicator is more politically sensitive than this one. At the international level, climbing food prices could lead to potentially unmanageable inflation, abrupt shifts in currency exchange rates, and widespread political instability. This, in turn, could jeopardize the security of investments in food-importing countries such as China, Egypt, and Mexico.

In the new era, political leaders will be called on to govern under unfamiliar conditions. Their understanding of the world, their values, and their priorities were shaped in a far different age. With the new era comes the need for different priorities in the use of public resources—priorities that recognize food scarcity rather than military aggression as the principal threat to security.

In an integrated economy where expanding human demand for food is colliding with the earth's natural limits, population growth anywhere limits the ability of people everywhere to consume more grain. Contrary to popular opinion, it will not be in the devastation of poverty-stricken Somalia or Haiti but in the booming economy of China that we will see the inevitable collision between the expanding demand for food and the limits of some of the earth's most basic natural systems.

In addition to raising food prices, the failure to arrest the deterioration of our basic life-support systems could bring economic growth to a halt, dropping incomes and food purchasing power throughout the world. It could lead to political unrest and a swelling flow of hungry migrants across national borders. Rising food prices and the associated economic and political disruptions within China could bring that nation's economic miracle to a premature end.

The new era calls for food carrying-capacity assessments on a country-by-country basis, much like the one done here for China. Existing information on cropland area, future availability of water for irrigation, and the potential contribution to grain production of existing agricultural technologies can provide the information governments need to project their own food production potential and measure it against their long-term needs. What many will discover is what Japan, South Korea, and Taiwan have learned: once population density reaches a certain point, further growth in numbers undermines the prospect for raising food consumption per person from indigenous resources. Countries lacking foreign exchange may find that continuous population growth will foreclose the options for diversifying diets and raising consumption.

The European Union, consisting of some 15 countries and containing 360 million people, provides a model for the rest of the world of an environmentally sustainable food/population balance. Europe is the first region to reach zero population growth. At the same time, movement up the food chain has also come to a halt as diets have become saturated with livestock products. The result is that Europe's grain consumption, which has not increased for close to a decade, has stabilized—and at a level that is within the region's carrying capacity. (See Figure 10–1.) Indeed, there is a potential for a small but sustainable export surplus of grain for the indefinite future.[2]

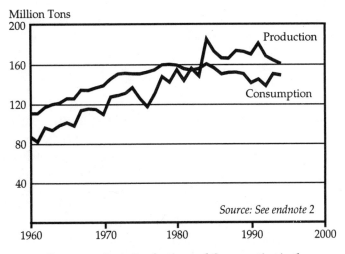

Million Tons

FIGURE 10-1. *Grain Production and Consumption in the European Community, 1960-94*

North America is the other region where grain consumption is currently below sustainable production. But the exportable surplus from these two regions will be less and less adequate to meet projected import needs elsewhere in the world. If countries cannot boost consumption per person from their indigenous resources, they may not be able to do so at all, given the likely competition for importable supplies.[3]

In the new era, by far the most urgent need is to stabilize world population as soon as possible. Based on national carrying capacity assessments, some countries may discover that the goals of the World Population Plan of Action adopted in Cairo in September 1994 are not sufficiently ambitious for them—that if they are to raise consumption levels, they may have to stabilize population size even sooner than envisaged in the plan.[4]

Closely parallelling the need to stabilize world population as soon as possible is the need to protect the resource base on which agriculture depends: soils, aquifers, and the climate system. In some agricultural regions, the thin layer of topsoil that accumulated over long periods of geological time is being gradually lost through erosion, undermining the inherent productivity of the land. In a world where the demand for food is beginning to press against the limits of supply, every ton of topsoil lost diminishes the food supply of the next generation. It is now clearly in the interest of the international community to devise a plan to stabilize soils—to reduce the loss from erosion below the rate of new soil formation through natural processes. In effect, there is a need for the agronomic equivalent of the World Population Plan of Action.

The depletion of aquifers by overpumping is a much more contemporary phenomenon than soil erosion, since it depends on quantities of energy for pumping that are available only in the modern era. Aquifer depletion is as undesirable and costly over the long term as it is widespread. Now common in the world's major food-producing regions, it is leading to falling water tables, higher pumping costs, and a misleading sense of food security, since a sizable fraction of today's harvest is based on the unsustainable use of water.[5]

The first priority is to determine the extent of aquifer depletion. But even while this is being done, the introduction of water marketing—pricing water at full cost—would reduce wasteful use and encourage investment in water-efficient technologies. Water marketing will also spur research that can lead to more water-efficient technologies and practices. Avoiding acute water scarcity depends on investing in water efficiency on a scale comparable to the investment in energy efficiency in the mid-seventies, and thus buying more time to stabilize population.

Conserving both soil and water depends on reversing the deforestation of the planet. If the goal is to reduce rainfall runoff, and thereby help increase aquifer recharge, and to stabilize soils, there is no alternative to planting trees. There is also a need to protect cropland from nonfarm uses. As noted for China, one of the principal threats to the world's cropland is the trend toward automobile-centered transportation systems: Not only does the evolution of an auto-centered transport system lead to the extensive paving of cropland, it also facilitates land-consuming urban sprawl.

Few things affect food production as much as changes in climate. World agriculture as it exists today is keyed to a climate system that has been remarkably stable for the last 10,000 years. Any changes in this will disrupt agriculture. The difficulties that the world's farmers are experiencing in trying to keep pace with population will intensify if they also have to cope with the disruptions of climate change.

With world carryover stocks of grain at their lowest level in 20 years and with the prospect of spreading food scarcity, an inventory is needed of the various reserves that can be tapped to alleviate scarcity and buy time to stabilize population. The most easily tapped reserve is the cropland in the United States and Europe that is idled under supply management programs for farm commodities, which are designed to avoid surpluses. (See Chapter 8.) If this land were returned to production, it could boost the world grain harvest by 2 percent, enough to cover the additional demand of world population growth for perhaps 15 months. If, in addition, the share of the U.S. Conservation Reserve Program land of 14 million hectares that could be farmed sustainably with appropriate soil-conserving farming practices were again plowed, it could conceivably boost the world grain harvest by enough to cover another 12 months of world population growth.[6]

Another source of land to produce food is the fields used to grow nonfood products, such as tobacco. If the 5 million hectares of cropland with tobacco growing on it were switched to grain, assuming the average world yield of 2.4 tons per hectare, it would provide enough grain to support the growth in world population growth for nearly six months.[7]

Almost as large a potential source of food is the 1.4 million hectares of highly productive U.S. cornland (8 tons per hectare) used to produce annually the 11 million tons of corn that in turn produce roughly 1 billion gallons of ethanol for use as an automotive fuel. Making this grain available for human consumption could cover four months of world population growth.[8]

The area planted to cotton could also be reduced. If consumers could be persuaded to replace half of the cotton clothing they buy with clothes made from synthetic fibers, some 9 million hectares of land would be freed up, providing enough grain for 11 months of world population growth. China, the world's leading cotton consumer, is already investing heavily in the manufacture of synthetic fibers on a scale that could eventually lower demand for cotton.[9]

Yet another source of additional grain is that which is wasted because of faulty storage or transportation. There are no good estimates on how much grain could realistically be saved or what improved storage would cost worldwide, but the gains are potentially much larger than some of those just cited. If grain prices rise, investment in improved grain storage facilities is likely to expand.

By far the largest food reserve is the 37 percent of the world grain harvest, some 630 million tons in 1994, that is used to produce livestock and poultry products for human consumption. This includes meat of various kinds, milk and milk products (butter, cheese, yogurt, and ice cream), eggs, and fish from aquaculture.[10]

To some degree, market forces will tap this reserve as rising grain prices push up prices of livestock products, reducing their consumption. Unfortunately, the price level at which a substantial reduction occurs is so high that it could force food consumption among millions of the world's poor below the survival level. Rationing the consumption of livestock products in the more affluent societies would free up grain without leading to dramatic price rises.

The same reduction in consumption could be achieved by imposing a tax on consumption of livestock products, one that would be similar to those that governments put on alcoholic beverages and cigarettes. Such a tax would affect the more affluent not only in industrial countries but in developing ones as well, since China is now the world's largest consumer of red meat.

If the 630 million tons of grain used for feed were reduced by 10 percent, by whatever means, it would free up 63 million tons of grain for consumption as food, enough to cover world population growth for 28 months. Unprecedented and unpopular though a livestock-products tax would be, it could be the price of global economic stability in an era of scarcity.[11]

Beyond this, an international food reserve is urgently needed—one that would acquire stocks when prices are low in order to release them when they are higher. In a world of food scarcity and soaring prices, the economic instability associated with inadequate reserves could lead to political turmoil and the downfall of governments. In an integrated world economy, political stability is essential to economic progress.

Overall, there is a need for much greater investment in agricultural research, although not because of the likelihood of another breakthrough like the development of hybrid corn or the discovery of chemical fertilizer that will lead to a huge gain in world food output. But in a world of food scarcity, every technological advance that helps expand production, however small, is important. Each one buys a little more time with which to stabilize population.

In the new era, achieving a balance between the supply and demand for food will increasingly depend on adjustments on the demand side. The bottom line is that achieving a humane balance between food and people is now more in the hands of family planners than farmers. For example, with oceanic fisheries now pushed to the limit, arresting the decline in the seafood catch per person now depends on stabilizing our numbers.

Time is not on our side. The world has waited too long to stabilize population. The decline in seafood supply per person and in grain output per person is already under way. This is not something that might happen. It is happening. Unfortunately, these trends are defining characteristics of the new era.

Effectively addressing these threats to our future will take a massive mobilization of resources, both financial and political. If we care about the future, we have no choice but to launch a worldwide effort to stabilize our life-support systems—soils, fisheries, aquifers, and forests—and the climate system. Historically, the only comparable mobilization of financial resources and political leadership was that during World War II. Without such an effort to deal with these new threats to our security, we will leave our children a world without hope. China's prospective emergence as a massive grain importer is a wake-up call—one that will force us to address issues we have long neglected.

Leaders are judged by whether or not they respond to the great issues of their time. For our generation, the overriding issue is whether we can reestablish a stable relationship between our numbers and aspirations on the one hand and the earth's natural support systems on the other. Unless we act quickly and decisively, neither history nor our children will judge us kindly.

Notes

CHAPTER 1. Overview: The Wake-Up Call

1. U.S. Department of Agriculture (USDA), Economic Research Service (ERS), "Production, Supply, and Demand View" (electronic database), Washington, D.C., November 1994.
2. Ibid.
3. Figure 1–1 from ibid., with 1950–59 figures from USDA, ERS, "World Grain Database" (unpublished printout), Washington, D.C., April 1989.
4. USDA, op. cit. 1; USDA, op. cit. note 3; U.S. Bureau of the Census, as published in Francis Urban and Ray Nightingale, *World Population by Country and Region, 1950–90 and Projections to 2050* (Washington, D.C.: USDA, ERS, 1993).
5. Patrick E. Tyler, "China Planning People's Car to Put Masses Behind Wheel," *New York Times*, September 22, 1994; Nicholas D. Kristof and Sheryl WuDunn, *China Wakes* (New York: Random House, 1994).
6. USDA, op. cit. note 1, with updates from USDA, ERS, "World Agriculture Production," Washington, D.C., March 1995.

7. Irrigated area from USDA, ERS, "China Situation and Outlook Series," Washington, D.C., July 1993, and from U.N. Food and Agriculture Organization (FAO), *Production Yearbooks* (Rome: various years); Professor Xu Zhifang, President, Chinese National Committee on Irrigation and Drainage, speech for the World Water Council-Interim Founding Committee, Montreal, Canada, March 31, 1995.

8. Figure 1–2 from USDA, op. cit. note 1, and from USDA, op. cit. note 3.

9. Figure 1–3 from USDA, op. cit. note 1, from USDA, op. cit. note 3, and from Bureau of the Census, op. cit. note 4.

10. Per capita figures from USDA, op. cit. note 1, and Bureau of the Census, op. cit. note 4.

11. Bureau of the Census, op. cit. note 4; population of Beijing from United Nations, *Estimates and Projections of Urban, Rural, and City Populations 1950–2025: The 1982 Assessment* (New York: 1985).

12. Kristof and WuDunn, op. cit. note 5; International Monetary Fund, *World Economic Outlook October 1994* (Washington, D.C.: 1994).

13. Bureau of the Census, op. cit. note 4.

14. FAO, *Food Balance Sheet 1988* (Rome: 1989).

15. USDA, op. cit. note 1; Bureau of the Census, op. cit. note 4.

16. Sheila Tefft, "A Shrinking Rice Bowl in China: Rising Food Prices Spur Unease," *Christian Science Monitor*, January 19, 1995; Gao Anming, "Experts Note Reasons for Hikes in Grain Prices," *China Daily*, January 6, 1995; Martin Wolf, "Zooming in on the Threat of Inflation," *Financial Times*, November 7, 1994; "China to Buy More Wheat, Corn From US, Traders Say," *Journal of Commerce*, January 19, 1995; "China Restricts Trading in Shanghai Rice Futures," *Journal of Commerce*, October 26, 1994; Joseph Kahn, "China Fails to Curb Its Runaway Growth," *Wall Street Journal*, January 3, 1995.

17. World Bank, *China: Strategies for Reducing Poverty in the 1990's* (Washington, D.C.: 1992).

18. Peter Hannam, "China Seen Facing Choice: Inflation or Unemployment," *Journal of Commerce*, September 27, 1994.

19. USDA, "Grain: World Markets and Trade," Washington, D.C., various issues.

20. Population Reference Bureau, *1994 World Population Data Sheet* (Washington, D.C.: 1994).

CHAPTER 2. Another Half-Billion

1. Figure 2–1 from U.S. Bureau of the Census, as published in Francis Urban and Ray Nightingale, *World Population by Country and Region, 1950–90 and Projections to 2050* (Washington, D.C.: U.S. Department of Agriculture (USDA), Economic Research Service (ERS), 1993); 1900 world population from United Nations, *The Future Growth of World Population* (Rome: 1958).

2. Bureau of the Census, op. cit. note 1.

3. Grainland per capita calculated with population from Bureau of the Census, op. cit. note 1, and grain area from USDA, ERS, "Production, Supply, and Demand View" (electronic database), Washington D.C., November 1994; grain trade data from ibid.

4. Population from Bureau of the Census, op. cit. note 1.

5. Susan Cotts Watkins and Jane Menken, "Famines in Historical Perspective," *Population and Development Review*, December 1985.

6. Ibid.

7. Figure 2–2 from United Nations, *Monthly Bulletin of Statistics*, February 1994.

8. Figure 2–3 from U.S. Bureau of the Census, Center for International Research, Suitland, Md., private communication, February 6, 1995.

9. John R. Bermingham, "China's Population Puzzle," Colorado Population Coalition, Denver, Colo., Fall 1994.

10. Michael S. Teitelbaum, "The Population Threat," *Foreign Affairs*, Winter 1992/93.

11. India's population from Bureau of the Census, op. cit. note 1.

12. Aaron Segal, *An Atlas of International Migration* (London: Hanz Zell Publishers, 1993).

13. Zha Ruichuan and Qiao Xiachun, "A Study of the Age Structure of China's Population," *China Population Today*, December 1992; Steve Mufson, "Chinese Leader Presses for 'One Couple, One Child,'" *Washington Post*, March 21, 1995.

14. Mufson, op. cit. note 13.

15. Population projection from Bureau of the Census, op. cit. note 1; Figure 2–4 from Zha and Qiao, op. cit. note 13.

16. Carl Haub, "China's Fertility Drop Lowers World Growth Rate," *Population Today*, June 1993.

17. Ibid.; Carl Haub, Population Reference Bureau, Washington, D.C., private communication, April 19, 1995.

18. Jiang quoted in Mufson, op. cit. note 13.

CHAPTER 3. Moving Up the Food Chain

1. Nicholas D. Kristoff, "Riddle of China: Repression as Standard of Living Soars," *New York Times*, September 7, 1993.

2. Population from Population Reference Bureau (PRB), *World Population Data Sheet 1994* (Washington, D.C.: 1994); economy from International Monetary Fund, *World Economic Outlook, October 1994* (Washington, D.C.: 1994).

3. U.N. Food and Agriculture Organization (FAO), *Food Balance Sheet 1985* (Rome: 1986).

4. Table 3–1 from FAO, *Food Production Yearbook 1993* (Rome: 1994).

5. Ibid.; energy ratio from United Nations, *Energy Statistics Yearbook 1992* (New York: 1994).

6. U.S. Department of Agriculture (USDA), Economic Research Service (ERS), "Livestock and Poultry: World Markets and Trade," Washington, D.C., October 1994.

7. USDA, ERS, "Production, Supply, and Demand View" (electronic database), Washington, D.C., November 1994, with updates from USDA, op. cit. note 6; meat consumption from FAO, op. cit. note 4.

8. Figure 3–1 from USDA, op. cit. note 7, with updates from USDA, op. cit. note 6.

9. USDA, op. cit. note 7; USDA, op. cit. note 6.

10. Grain-to-poultry ratio derived from Robert V. Bishop et al., *The World Poultry Market—Government Intervention and Multilateral Policy Reform* (Washington, D.C.: USDA, 1990); grain-to-pork ratio from Leland Southard, Livestock and Poultry Situation and Outlook Staff, ERS, USDA, Washington, D.C., private communication, April 27, 1992; grain-to-beef ratio based on Allen Baker, Feed Situation and Outlook Staff, ERS, USDA, Washington, D.C., private communication, April 27, 1992.

11. USDA, op. cit. note 6; USDA, ERS, "China Situation and Outlook Series," Washington, D.C., July 1993.

12. USDA, op. cit. note 6; grain-to-egg conversion ratio from Alan B. Durning and Holly B. Brough, *Taking Stock: Animal Farming and the Environment*, Worldwatch Paper 103 (Washington, D.C.: Worldwatch Institute, July 1991), citing USDA, Foreign Agricultural Service, *World Livestock Situation*, Washington, D.C., April 1991, and Linda Baily, agricultural economist, USDA, Washington, D.C., private communication, September 11, 1990.

13. Beef trade from USDA, op. cit. note 6; depletion of rangelands from Lester R. Brown and Hal Kane, *Full House: Reassessing the*

Earth's Population Carrying Capacity (New York: W.W. Nɔrton & Company, 1994).

14. FAO, op. cit. note 4.
15. FAO, *Yearbook of Fishery Statistics: Catches and Landings* (Rome: 1993); population from PRB, op. cit. note 2.
16. FAO, "Marine Fisheries and the Law of the Sea: A Decade of Change," Fisheries Circular No. 853, Rome, 1993.
17. FAO, "Aquaculture Production, 1985–91," Rome, 1992; grain-to-fish conversion ratio from Ross Garnaut and Guonan Ma, East Asian Analytical Unit, Department of Foreign Affairs and Trade, *Grain in China* (Canberra: Australian Government Publishing Service, 1992).
18. Figure 3–2 from USDA, op. cit. note 7.
19. "Americans Find China a Heady Brew," *Financial Times*, March 28, 1995.
20. Grain-to-beer ratio from Jack McCabe, Brew Master, Chicago, Ill., private communication, June 10, 1994, and from Virginia Brewers Association, private communication, June 12, 1994; USDA, op. cit. note 7; FAO, "Time Series for State of Food and Agriculture" (electronic database), Rome, May 1994; "Americans Find China a Heady Brew," op. cit. note 19.
21. FAO, *Food Balance Sheet 1988* (Rome: 1989); USDA, "Oilseeds Situation and Outlook," Washington, D.C., July 1994; populations from PRB, op. cit. note 2.
22. "Production Shortfall May Force China to Increase Soy Oil Imports From U.S.," *Journal of Commerce*, March 8, 1995; Figure 3–3 from USDA, op. cit. note 7.
23. FAO, op. cit. note 21.

CHAPTER 4. The Shrinking Cropland Base

1. "Chinese Reform Burial Customs," *Mazingira*, March 1984.
2. U.N. Food and Agriculture Organization, *1994 Production Yearbook* (Rome: 1993).
3. U.S. Department of Agriculture (USDA), Economic Research Service (ERS), "Production, Supply, and Demand View" (electronic database), Washington, D.C., November 1994; population from U.S. Bureau of the Census, as published in Francis Urban and Ray Nightingale, *World Population by Country and Region, 1950–90 and Projections to 2050* (Washington, D.C.: USDA, ERS, 1993).
4. Zou and *Economic Information Daily* from Tony Walker, "China Determined to Head Off Farmland Crisis," *Financial Times*, March 2, 1995.
5. Figure 4–1 from USDA, op. cit. note 3, with 1950–59 data from

USDA, "World Grain Database" (unpublished printout), Washington, D.C., April 1989.

6. Population projections from Bureau of the Census, op. cit. note 3.

7. David Malin Roodman and Nicholas Lenssen, *A Building Revolution: How Ecology and Health Concerns Are Transforming Construction*, Worldwatch Paper 124 (Washington, D.C.: Worldwatch Institute, March 1995).

8. Joseph Kahn, "China's Next Great Leap: The Family Car," *Wall Street Journal*, June 24, 1994; Sun Shangwu, "Building Eats Up Farmland as More Mouths Need Food," *China Daily*, July 18, 1994.

9. "Chinese Roads Paved With Gold," *Financial Times*, November 23, 1994.

10. Figure 4–2 from ibid.

11. P.T. Bangsberg, "China Presses On With Expansion of Highway System," *Journal of Commerce*, January 18, 1995.

12. Ibid.; Patrick E. Tyler, "Hong Kong Tycoon's Road to China," *New York Times*, December 31, 1993.

13. Wang Rong, "Food Before Golf on Southern Land," *China Daily*, January 25, 1995.

14. World Bank, *China: Strategies for Reducing Poverty in the 1990's* (Washington, D.C.: 1992).

15. Ministry of Agriculture, Forestry, and Fisheries, *Statistical Yearbook of Agriculture, Forestry and Fisheries* (Tokyo: various years); Ministry of Agriculture, Forestry and Fisheries, *Statistical Yearbook of Agriculture, Forestry and Fisheries* (Seoul: various years); Taiwan data from John Dyck, USDA, Foreign Agricultural Service, Washington, D.C., private communication, March 16, 1995.

16. Figure 4–3 from Ministry of Agriculture, Forestry, and Fisheries (Tokyo), op. cit. note 15, from Ministry of Agriculture, Forestry and Fisheries (Seoul), op. cit. note 15, and from Dyck, op. cit. note 15.

17. W. Hunter Colby et al., *Agricultural Statistics of the People's Republic of China, 1949–90* (Washington, D.C.: USDA, ERS, 1992), with updates from Dyck, op. cit. note 15.

18. Table 4–1 from USDA, ERS, "China Situation and Outlook Report," Washington, D.C., August 1994.

19. Figure 4–4 from Worldwatch, with projections based on USDA, op. cit. note 3, and on Bureau of the Census, op. cit. note 3, assuming current rate of cropland loss.

20. Noel Grove, "Rice, the Essential Harvest," *National Geographic*, May 1994; USDA, op. cit. note 3; USDA, op. cit. note 5.

CHAPTER 5. Spreading Water Scarcity

1. Population from U.S. Bureau of the Census, as published in Francis Urban and Ray Nightingale, *World Population by Country and Region, 1950–90 and Projections to 2050* (Washington, D.C.: U.S. Department of Agriculture (USDA), Economic Research Service (ERS), 1993); James Nickum and John Dixon, "Environmental Problems and Economic Modernization," *Asia Pacific Report, Focus: China in the Reform Era* (Honolulu, Hawaii: East-West Center, 1989).
2. Niu quoted in Patrick E. Tyler, "China Lacks Water to Meet Its Mighty Thirst," *New York Times*, November 7, 1993.
3. Xu Zhifang, President, Chinese National Committee on Irrigation and Drainage, speech at the World Water Council-Interim Founding Committee, Montreal, Canada, March 31, 1995; U.N. Food and Agriculture Organization (FAO), *Production Yearbook* (Rome: various years).
4. Patrick E. Tyler, "Huge Water Project Would Supply Beijing By 860-Mile Aqueduct," *New York Times*, July 19, 1994.
5. Figure 5–1 from USDA, ERS, "China Situation and Outlook Series," Washington, D.C., August 1994, and from FAO, op. cit. note 3; James E. Nickum, "Volatile Waters: Is China's Irrigation in Decline?" Environment and Policy Institute, East-West Center, presented at the 81st Annual Meeting of the American Society of Agronomy, Las Vegas, Nev., October 19, 1989.
6. USDA, op. cit. note 5; FAO, op. cit. note 3; population from Bureau of the Census, op. cit note 1.
7. USDA, op. cit. note 5; FAO, op. cit. note 3; W. Hunter Colby, Frederick W. Crook, and Shwu-Eng H. Webb, *Agricultural Statistics of the People's Republic of China, 1949–90* (Washington, D.C.: USDA, ERS, 1992).
8. Colby, Crook, and Webb, op. cit. note 7; Tyler, op. cit. note 4.
9. Colby, Crook, and Webb, op. cit. note 7; USDA, op. cit. note 5.
10. Vaclav Smil, *China's Environmental Crisis: An Inquiry Into the Limits of National Development* (Armonk, N.Y.: M.E. Sharpe, 1993).
11. Ibid.
12. Ibid.
13. Tyler, op. cit. note 4.
14. Ibid.
15. Smil, op. cit. note 10; Xu, op. cit. note 3.
16. Harald D. Frederiksen, Jeremy Berkoff, and William Barber,

"Water Resources Management in Asia, Volume 1," World Bank Technical Paper 212, Washington, D.C., 1993.

17. Bureau of the Census, op. cit note 1.
18. World Bank, *World Development Report 1994* (New York: Oxford University Press, 1994); Nickum and Dixon, op. cit. note 1.
19. Bureau of the Census, op. cit. note 1; Nickum and Dixon, op. cit. note 1.
20. Tyler, op. cit. note 4.
21. Colby, Crook, and Webb, op. cit. note 7.
22. Tyler, op. cit. note 4.
23. Ibid.
24. Ibid.
25. Xu, op. cit. note 3.

CHAPTER 6. Raising Cropland Productivity

1. U.S. Department of Agriculture (USDA), Economic Research Service (ERS), "Production, Supply, and Demand View" (electronic database), Washington, D.C., November 1994.
2. Ibid.
3. Ibid.; data for 1950–59 from USDA, ERS, "World Grain Database" (unpublished printout), Washington, D.C., April 1989; Susan Cotts Watkins and Jane Menken, "Famines in Historical Perspective," *Population and Development Review*, December 1985.
4. USDA, op. cit. note 1; USDA, op. cit. note 3.
5. U.N. Food and Agriculture Organization (FAO), *Fertilizer Yearbooks* (Rome: various years); USDA, op. cit. note 1.
6. 1880 data from Japanese Ministry of Agriculture, Forestry and Fisheries, *Crop and Livestock Statistics*, various years; Figure 6–1 from USDA, op. cit. note 1, from USDA, op. cit. note 3, and from USDA, ERS, "World Agricultural Production," Washington, D.C., February 1995; rice prices from USDA, ERS, "Pacific Rim Agriculture and Trade Report, Situation and Outlook Series," Washington, D.C., September 1992.
7. USDA, op. cit. note 1.
8. Keith Schneider, "A New Rice Could Raise Yields 20%," *New York Times*, October 24, 1994.
9. Figure 6–2 from USDA, op. cit. note 1, from USDA, op. cit. note 3, and from USDA, "World Agricultural Production," op. cit. note 6.
10. USDA, op. cit. note 1.
11. Ibid.; USDA, "World Agricultural Production," op. cit. note 6.
12. Table 6–1 from USDA, op. cit. note 1, from USDA, op. cit.

note 3, and from USDA, "World Agricultural Production," op. cit. note 6.

13. USDA, op. cit. note 1; USDA, "China Situation and Outlook Series," Washington, D.C., August 1994; FAO, op. cit. note 5.

14. USDA, op. cit. note 1.

15. Ibid.; USDA, op. cit. note 3; Schneider, op. cit. note 8.

16. Donald N. Duvick, "Intensification of Known Technology and Prospects of Breakthroughs in Technology and Future Food Supply," Iowa State University, Johnstown, Iowa, February 1994.

17. Figure 6–3 from FAO, op. cit. note 5, and from K.F. Isherwood and K.G. Soh, "The Agricultural Situation and Fertilizer Demand," presented at 62nd Annual Conference, International Fertilizer Industry Association, Istanbul, May 9, 1994.

18. FAO, op. cit. note 5; Isherwood and Soh, op. cit. note 17.

19. FAO, op. cit. note 5; Isherwood and Soh, op. cit. note 17.

20. Isherwood and Soh, op. cit. note 17; USDA, op. cit. note 1.

21. "Efforts Reduce Erosion Along the Yellow River," *China Daily*, January 4, 1995; Josef R. Parrington et al., "Asian Dust: Seasonal Transport to the Hawaiian Islands," *Science*, April 8, 1983.

22. Lester R. Brown and Jodi L. Jacobson, *The Future of Urbanization: Facing the Ecological and Economic Constraints*, Worldwatch Paper 77 (Washington, D.C.: Worldwatch Institute, May 1987).

23. Marcus W. Brauchli, "China's Environment Is Severely Stressed As Its Industry Surges," *Wall Street Journal*, July 25, 1994; James J. MacKenzie and Mohammed T. El-Ashry, *Ill Winds: Airborne Pollution's Toll on Trees and Crops* (Washington, D.C.: World Resources Institute, 1988).

CHAPTER 7. The Growing Grain Deficit

1. Nikos Alexandratos, "The Outlook for World Food and Agriculture to the Year 2010," U.N. Food and Agriculture Organization (FAO), Rome, January 1994; Donald O. Mitchell and Merlinda D. Ingco, International Economics Department, *The World Food Outlook* (Washington, D.C.: World Bank, November 1993); Aditi Kapoor, "China's Food Shortage Will Hit World Economy: World Watch," *Times of India*, February 2, 1995.

2. Per capita grainland calculated using grain area from U.S. Department of Agriculture (USDA), Economic Research Service (ERS), "Production, Supply, and Demand View" (electronic

database), Washington, D.C., November 1994, with 1950–59 figures from USDA, ERS, "World Grain Database" (unpublished printout), Washington, D.C., April 1989, and population from the U.S. Bureau of the Census, as published in Francis Urban and Ray Nightingale, *World Population by Country and Region, 1950–90 and Projections to 2050* (Washington, D.C.: USDA, ERS, 1993).

3. Figure 7–1 from USDA (electronic database), op. cit. note 2, and from USDA (unpublished printout), op. cit. note 2.

4. USDA (electronic database), op. cit. note 2; USDA (unpublished printout), op. cit. note 2; USDA, ERS, "Grain: World Markets and Trade," Washington, D.C., March 1995.

5. Figure 7–2 from USDA (electronic database), op. cit. note 2, from USDA (unpublished printout), op. cit. note 2, and from USDA, op. cit. note 4.

6. USDA (electronic database), op. cit. note 2; USDA (unpublished printout), op. cit. note 2; USDA, op. cit. note 4.

7. USDA (electronic database), op. cit. note 2; USDA (unpublished printout), op. cit. note 2; USDA, op. cit. note 4.

8. Figure 7–3 from USDA (electronic database), op. cit. note 2, from USDA (unpublished printout), op. cit. note 2, and from USDA, op. cit. note 4.

9. Meat consumption from USDA (electronic database), op. cit. note 2; USDA, "Livestock and Poultry: World Markets and Trade," Washington, D.C., October 1994.

10. USDA (electronic database), op. cit. note 2; USDA, op. cit. note 4.

11. Seafood consumption from FAO, *Fishery Statistics: Catches and Landings* (Rome: 1993).

12. Ibid.; grain-fish conversion from Ross Garnaut and Guonan Ma, East Asian Analytical Unit, Department of Foreign Affairs and Trade, *Grain in China* (Canberra: Australian Government Publishing Service, 1992); USDA, op. cit. note 4.

13. Xu Zhifang, President, Chinese National Committee on Irrigation and Drainage, speech at the World Water Council-Interim Founding Committee, Montreal, Canada, March 31, 1995.

14. Keith Schneider, "A New Rice Could Raise Yields 20%," *New York Times*, October 24, 1994; "WASDE Cotton US & World," *Ag News Fax,* USDA, Washington, D.C., May 10, 1995.

15. USDA (electronic database), op. cit. note 2; USDA, op. cit. note 4.

16. Per capita consumption calculated from USDA (electronic database), op. cit. note 2, and from population from Bureau of the Census, op. cit. note 2.

17. Per capita consumption calculated from USDA (electronic database), op. cit. note 2, and from population from Bureau of the Census, op. cit. note 2.

18. Figure 7–4 is from Worldwatch, based on USDA (electronic database), op. cit. note 2, and on Bureau of the Census, op. cit. note 2.

19. Figure 7–5 is from Worldwatch, based on USDA (electronic database), op. cit. note 2, and on Bureau of the Census, op. cit. note 2.

20. Populations from Bureau of the Census, op. cit. note 2; grain trade from USDA (electronic database), op. cit. note 2.

21. Zhou quoted in Patrick E. Tyler, "The Dynamic New China Still Races Against Time," *New York Times*, January 2, 1994.

22. Population Reference Bureau, *World Population Data Sheet 1994* (Washington, D.C.: 1994).

23. "China to Buy More Wheat, Corn From US, Traders Say," *Journal of Commerce*, January 19, 1995; International Monetary Fund, *World Economic Outlook, October 1994* (Washington, D.C.: 1994).

24. Gao Anming, "Experts Note Reasons for Hikes in Grain Prices," *China Daily*, January 6, 1995.

25. "China to Buy More Wheat," op. cit. note 23; "China Expects to Control Inflation By Imposing a Ban on Corn Exports," *Journal of Commerce*, December 5, 1994.

26. USDA, op. cit. note 4.

27. Peter Hannam, "China's Grain Shortages Echo Experts Warnings," *Journal of Commerce*, April 25, 1995.

CHAPTER 8. Competition for Grain

1. Joseph Kahn, "Creating a Wonder for China: Family Car," *Wall Street Journal*, November 21, 1994.

2. International Monetary Fund (IMF), *International Financial Statistics Yearbook* (Washington, D.C.: 1994).

3. Table 8–1 from International Trade Administration, "Total Imports and Exports for Individual Countries, 1980–93" (electronic database), National Trade Data Bank (NATB), Washington, D.C., 1994, with updates from U.S. Bureau of the Census, "U.S. Exports to China," and "U.S. Imports from China," NATB, op. cit. this note; price of corn from IMF, op. cit. note 2.

4. U.S. Department of Agriculture (USDA), Economic Research Service (ERS), "Production, Supply, and Demand View" (electronic database), Washington, D.C., November 1994.

5. Table 8–2 from ibid., with historical data from USDA, ERS, "World Grain Database" (unpublished printout), Washington, D.C., April 1989, and from U.N. Food and Agriculture Organization (FAO), *Trade Yearbook* (Rome: various years).

6. USDA. op. cit. note 4; USDA, op. cit. note 5; FAO, op. cit. note 5.

7. USDA. op. cit. note 4; USDA, op. cit. note 5; FAO, op. cit. note 5.

8. USDA, op. cit. note 4; USDA, op. cit. note 5; USDA, "Grain: World Markets and Trade," Washington, D.C., March 1995.

9. USDA, ERS, "Oilseeds Situation and Outlook," Washington, D.C., January 1995.

10. USDA, op. cit. note 4.

11. Ibid.

12. USDA, ERS, "Grain Situation and Outlook Report," Washington, D.C., December 1994.

13. Population growth rates from U.S. Bureau of the Census, as published in Francis Urban and Ray Nightingale, *World Population by Country and Region, 1950–90 and Projections to 2050* (Washington, D.C.: USDA, ERS, 1993).

14. USDA, op. cit. note 4.

15. USDA, op. cit. note 12; USDA, op. cit. note 4.

16. USDA, op. cit. note 12; USDA, op. cit. note 4.

17. Gordon Sloggett and Clifford Dickason, *Ground-Water Mining in the United States* (Washington, D.C.: USDA, ERS, 1986).

18. Figure 8–1 from USDA, ERS, *Agricultural Statistics* (Washington, D.C.: various years).

19. Population projections from Bureau of the Census, op. cit. note 13.

20. Grain consumption from USDA, op. cit. note 4; population from Bureau of the Census, op. cit. note 13.

21. USDA, "World Agriculture Production," Washington, D.C., various issues; stocks from USDA, op. cit. note 4.

22. Grain stocks from USDA, op. cit. note 4; James Hansen, "The Greenhouse Effect: Impacts on Current Global Temperature and Regional Heat Waves," statement before the Committee on Energy and Natural Resources, U.S. Senate, Washington, D.C., June 23, 1988, and as quoted in Michael Weisskopf, " 'Greenhouse Effect' Fueling Policy Makers," *Washington Post*, August 15, 1988.

23. Figure 8–2 from USDA, op. cit. note 4, from USDA, op. cit. note 5, and from FAO, op. cit. note 5.

24. USDA, op. cit. note 4.

25. Ibid.

26. Ibid.
27. Suzanne McGee, "Sugar Soars in Rumors That China Might Buy," *Wall Street Journal*, November 23, 1994.
28. USDA, op. cit. note 9; USDA, ERS, "Agricultural Outlook," Washington, D.C., May 1995.
29. Table 8-3 contains Worldwatch estimates based on USDA, op. cit. note 4, and on population projections from Bureau of the Census, op. cit. note 13.
30. Lester R. Brown and Hal Kane, *Full House: Reassessing the Earth's Population Carrying Capacity* (New York: W.W. Norton & Company, 1994).
31. Ibid.; wheat prices from IMF, op. cit. note 2.
32. Figure 8-3 from World Bank, *Commodity Trade and Price Trends 1989–91 Edition* (Baltimore, Md.: Johns Hopkins University Press, 1993).

CHAPTER 9. Entering a New Era

1. Population increase from U.S. Bureau of the Census, as published in Francis Urban and Ray Nightingale, *World Population by Country and Region, 1950–90 and Projections to 2050* (Washington, D.C.: U.S. Department of Agriculture (USDA), Economic Research Service (ERS), 1993); economic expansion from International Monetary Fund (IMF), *International Monetary Statistics Yearbook* (Washington, D.C.: 1994).
2. Aditi Kapoor, "China's Food Shortage Will Hit World Economy: World Watch," *Times of Bombay*, February 2, 1995; grain yields from USDA, ERS, "Production, Supply, and Demand View" (electronic database), Washington, D.C., November 1994.
3. Population and projections from Bureau of the Census, op. cit. note 1; John R. Bermingham, "China's Population Puzzle," Colorado Population Coalition, Denver, Colo., Fall 1994.
4. Xu Zhifang, President, Chinese National Committee on Irrigation and Drainage, speech for the World Water Council-Interim Founding Committee, Montreal, Canada, March 31, 1995.
5. Patrick E. Tyler, "Nature and Economic Boom Devouring China's Farmland," *New York Times*, March 27, 1994.
6. Nikos Alexandratos, "The Outlook for World Food and Agriculture to the Year 2010," U.N. Food and Agriculture Organization (FAO), Rome, January 1994; Donald O. Mitchell and Merlinda D. Ingco, International Economics Department, *The World Food Outlook* (Washington, D.C.: World Bank, November 1993).

7. Table 9–1 is from FAO, *Fishery Statistics: Annual Catches and Landings* (Rome: various years), from USDA, op. cit. note 2, and from USDA, "World Grain Database" (unpublished printout), Washington, D.C., April 1989.

8. USDA, op. cit. note 2; USDA, op. cit. note 7.

9. USDA, op. cit. note 2.

10. Lester R. Brown, Nicholas Lenssen, and Hal Kane, *Vital Signs 1995* (New York: W.W. Norton & Company, 1995), as adapted from USDA, op. cit. note 2, and from USDA, "Grain: World Markets and Trade," Washington, D.C., January 1995.

11. Table 9–2 is from Worldwatch, based on FAO, op. cit. note 6, on USDA, op. cit. note 2, on Bureau of the Census, op. cit. note 1, and on U.N. Development Programme (UNDP), *Human Development Report 1994* (New York: Oxford University Press, 1994); rice harvest from USDA, op. cit. note 2.

12. Seafood prices from FAO, *Fishery Statistics: Trade and Commerce* (Rome: various years), with updates from Adele Crispoldi, fishery statistician, Fishery Information, Data and Statistics Service, Fisheries Department, FAO, Rome, unpublished printout, September 12, 1994.

13. Table 9–3 is Worldwatch estimates, based on World Bank, unpublished printout, February 1992, on gross world product data for 1950 and 1955 from Herbert R. Block, *The Planetary Product for 1980: A Creative Pause?* (Washington, D.C.: U.S. Department of State, 1981), on U.S. Bureau of the Census, Center for International Research, Suitland, Md., private communication, March 26, 1993, and on IMF, *World Economic Outlook: Interim Assessment* (Washington, D.C.: 1993).

14. IMF, op. cit. note 1.

15. UNDP, op. cit. note 11.

CHAPTER 10. Priorities in an Era of Scarcity

1. U.S. Bureau of the Census, as published in Francis Urban and Ray Nightingale, *World Population by Country and Region, 1950–90 and Projections to 2050* (Washington, D.C.: U.S. Department of Agriculture (USDA), Economic Research Service (ERS), 1993); USDA, ERS, "Production, Supply, and Demand View" (electronic database), Washington, D.C., November 1994; USDA, "World Grain Database" (unpublished printout), Washington, D.C., April 1989.

2. Figure 10–1 from Bureau of the Census, op. cit. note 1, from USDA (electronic database), op. cit. note 1, and from USDA (unpublished printout), op. cit. note 1.

3. USDA (electronic database), op. cit. note 1.

4. U.N. General Assembly, "Programme of Action of the United Nations International Conference on Population and Development" (draft), New York, September 19, 1994.

5. Sandra Postel, *Last Oasis: Facing Water Scarcity* (New York: W.W. Norton & Company, 1992).

6. USDA, ERS, "Grain: World Markets and Trade," Washington, D.C., March 1995.

7. USDA (electronic database), op. cit note 1; Bureau of the Census, op. cit. note 1.

8. USDA (electronic database), op. cit note 1; Bureau of the Census, op. cit. note 1.

9. USDA (electronic database), op. cit. note 1.

10. Ibid.

11. Consumption from ibid.

Index

ABOUT THE AUTHOR

LESTER R. BROWN is President of the Worldwatch Institute, a private, nonprofit environmental research organization in Washington, D.C. He is the recipient of a MacArthur Foundation "genius award," the United Nations' 1989 environment prize, and the Asahi Glass Foundation's Blue Planet Prize, and he holds a string of honorary degrees from universities around the world. The Library of Congress has requested Mr. Brown's personal papers and manuscripts, recognizing his role in shaping the global environmental movement. Before founding Worldwatch, he was Administrator of the U.S. Department of Agriculture's International Agricultural Development Service and Advisor to the Secretary. He holds degrees from Rutgers University, the University of Maryland, and Harvard University. Mr. Brown started his career as a farmer, growing tomatoes in southern New Jersey.